Seeing Is Believing

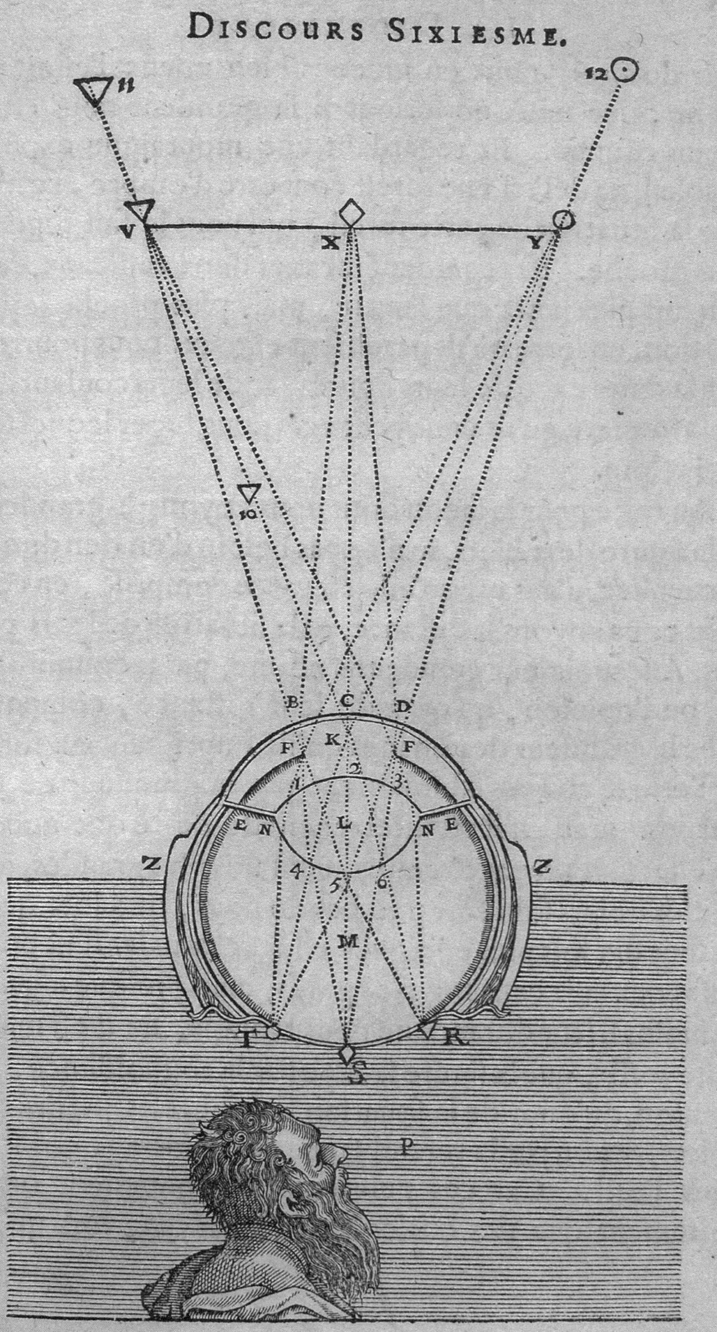
11
12
V
X
Y
10
B
C
D
F
K
F
2
1
3
E N
L
N E
Z
Z
4
5
6
M
T
R
S
P

700 Years of Scientific and Medical Illustration

Seeing Is Believing

JENNIFER B. LEE

and

MIRIAM MANDELBAUM

The New York Public Library

1999

Published for the exhibition
Seeing Is Believing: 700 Years of Scientific and Medical Illustration
presented at The New York Public Library
Humanities and Social Sciences Library
D. Samuel and Jeane H. Gottesman Exhibition Hall
October 23, 1999–February 19, 2000

Bibliographic information for items illustrated in this book may be found in the exhibition checklist, on pp. 62–74.

Lee, Jennifer B.
Seeing is believing : 700 years of scientific and medical illustration / Jennifer B. Lee and Miriam Mandelbaum.
p. cm.
Includes bibliographical references.
ISBN 0-87104-449-8
1. Scientific illustration—Exhibitions. 2. Medical illustration—Exhibitions. 3. New York Public Library—Exhibitions. I. Mandelbaum, Miriam, 1950– . II. Title.

Q222.L44 1999
502.2—dc21 99-046587

Printed on acid-free paper
Printed in the United States of America

EDITED BY BARBARA BERGERON

DESIGNED BY ANN ANTOSHAK

FRONTISPIECE: In each of the three works published as part of his 1637 *Discours de la méthode* (*La géométrie, La dioptrique*, and *Les météores*), René Descartes relied on experimentation, observation, and laborious mathematical calculation. This illustration from the work on light, *La dioptrique*, shows a cross section of the human eye and a demonstration of the way light is perceived.

Contents

Preface

As Leonhart Fuchs noted in the introduction to his great herbal of 1542, pictures "can communicate information much more clearly than the words of even the most eloquent men." In our own century, Ronald N. Giere, a professor of philosophy, has suggested that scientific theory is more like a picture than anything that can be captured in words. Indeed, scientific and medical illustration often allows a reader to "see" information that cannot actually be seen, by using different methods to show various kinds of theory or reality. Images such as Copernicus's simple diagram of the solar system and Gilbert's diagram of the polar North present theory based on careful study. Other images, such as Vesalius's elegant depictions of the muscles of the human body, and Trouvelot's glorious attempts to capture the wonders of the heavens, are based on observed, though selective, reality. A third type of image shows the way to conduct an experiment or procedure, or simply the equipment needed. Examples include Dürer's use of perspective in drawing a lute, Apianus's astronomical devices, and the equipment Boyle used in his experiments on air.

Seeing Is Believing: 700 Years of Scientific and Medical Illustration, published on the occasion of a major exhibition at The New York Public Library's Humanities and Social Sciences Library, presents a selection of scientific and medical illustrations dating from the thirteenth century through the beginning of the twentieth century, drawn primarily from the collections of illustrated books of science and medicine in the

Library's four research centers, augmented by materials from The New York Academy of Medicine and from a private collector. Although not providing a comprehensive history of scientific and medical illustration, these images open a window on the radical shift in the cosmology of early modern Europe that began around 1543 with the publication of seminal works by Copernicus and Vesalius, and continued with the work of Newton, Harvey, Darwin, Curie, and others.

Today scientists continue to develop and use new tools to break the bounds of human perception by exploring ever smaller subatomic realms and by searching ever farther away in the universe for its origins and for other forms of life. The past hundred years, in particular, have been witness to tremendous changes in the sciences and medicine and warrant separate attention elsewhere. From the vantage point of the new millennium, *Seeing Is Believing* calls attention to primarily Western science of earlier periods and to the artistry and skill that gave visual expression to those ideas.

Jennifer B. Lee
Associate Curator of Rare Books

Miriam Mandelbaum
Rare Books Librarian

Rare Books Division, Humanities and Social Sciences Library
The New York Public Library

mit einem anderen puncten aber also piß das du die gantzen lauten gar an die tafel punctirst / dann zeuch all puncten die auf der tafel von der lauten worden sind mit linien zůsamē/so sichst du was darauß wirt/also magst du ander ding auch abzeychnen. Dise meynung hab jch hernach aufgerissen.

Vnd damit günstiger lieber Herr will jch meinem schreyben end geben / vnd so mir Got genad verleycht die bůcher so jch von menschlicher proporcion vñ anderen darzů gehörend geschryben hab mit der zeyt in druck pringen/vnd darpey meniglich gewarnet haben/ob sich yemand vndersteen wurd mir diß außgangen bůchlein wider nach zů drucken /das jch das selb auch wider drucken will / vñ auß lassen geen mit meren vnd grösserem zůsatz dañ ietz beschehen ist/darnach mag sich ein yetlicher richtē/Got dem Herren sey lob vnd eer ewigklich.

Gedruckt zů Nůremberg.
Im. 1525. Jar.

About The New York Public Library's Science and Medicine Collections

At its founding in 1895, The New York Public Library already possessed a splendid array of important books in the fields of science and medicine. These came to the new library from the two private collections whose merger, along with a bequest from the Tilden Trust, created the new institution.

The first of those private collections, the Astor Library, founded in 1848 through the bequest of John Jacob Astor, was very strong in first and early editions of astronomy, mathematics, physics, chemistry, medicine, natural history, and microscopy. The Astor Library included such great medical works as the extremely rare first edition of William Harvey's landmark treatise on the circulation of the blood (1628) and William Hunter's work on the gravid uterus (1774). In the sciences, it included the first edition with the rare errata sheet of Nicolaus Copernicus's *De revolutionibus orbium coelestium* (1543); Robert Hooke's

The first of three works on technical and scientific subjects published by Albrecht Dürer toward the end of his life, the *Unterweysung der Messung* (1525) shows him to have been not only a great artist but a great mathematician. To bring to northern European artists, architects, and other craftsmen the rules of geometry and the science of perspective developed in Italy during the Renaissance, he invented a number of ingenious devices that he called "perspective apparatus" to assist the artist in the science of projective geometry. Dürer's illustrations demonstrate his own superb mastery of the art and science of perspective, showing the illusion of depth on a flat plane, in this case while drawing a lute.

Micrographia (1665); and a full set of Leopold Trouvelot's magnificent *Astronomical Drawings* (1882), to name just a few of the more than thirty Astor Library items included in the exhibition *Seeing Is Believing*.

The Lenox Library, founded in 1876 through the bequest of James Lenox, included many important books of science and medicine, as well as many books on natural history from the collection of Robert Leighton Stuart, which had become part of the Lenox Library in 1892. The Lenox Library owned not only the original elephant folio edition of Audubon's *Birds of America* (1827–38) but also a full set of the never-completed American reprint (1860–61), made by Julius Bien using the process of chromolithography. It had been given to the Lenox Library by Anna Palmer Draper, who would be a major supporter of The New York Public Library in its early years. The Stuart Collection included a copy of the first English edition of Euclid's *Elements* (1570) with the "Mathematical Preface" by John Dee, and a copy of Edward Lear's magnificent *Illustrations of the Family of Psittacidae, or Parrots* (1832), the first illustrated work of ornithology devoted to a single family of birds.

Dr. John Shaw Billings, the first Director of The New York Public Library, was also a physician of considerable stature: he had been Assistant Surgeon General of the United States and head of the Johns Hopkins Medical School prior to his appointment at the Library. He was also a friend and colleague of Sir William Osler, the noted physician and bookman who generously bestowed first editions of Andreas Vesalius's *De humani corporis fabrica* (1543) on various libraries, including the Library of Congress and The New York Academy of Medicine. (It is conjectured that Osler did not donate a copy to The New York Public Library because he knew that the Academy's copy would be available to the general public.) The first Vesalius *Fabrica* to come to The New York Public Library was thus a second folio edition (1555), which came as part of the original bequest that formed the Henry W. and Albert A. Berg Collection of English and American Literature. The Berg brothers, whose principal collecting interest was nineteenth-century British and American literature, were both prominent New York physicians, and no self-respecting physician–book collector would have been without a Vesalius. The Bergs were wide-ranging in their definition of literature; in addition to the works of Charles Dickens, they collected, for instance, the works of Charles Darwin, including *On the Origin of Species* (1859) and *The Descent of Man* (1871).

The second important *Fabrica* to come to the Library was purchased in the 1930s. The Library's interest in printing history and the private press movement made mandatory the purchase for the collections of the 1935 Bremer Presse edition, made from the original woodblocks (which would not survive World War II) and published by the University of Munich in collaboration with The New York Academy of Medicine, an institution with which The New York Public Library has always had a special relationship. Because the Academy maintains a medical research library that is open to the public – the only medical library in New York City that affords such access – The New York Public Library does not seek to duplicate its resources, and therefore collects medicine only in cases where a medical book complies in some other way with one of the Library's collecting policies, such as printing history, the African American experience, and medicine in art and music.

After the death of John Shaw Billings in 1913, Anna Palmer Draper made provisions for a special fund in his memory. When she died the following year, her bequest to The New York Public Library included not only her own books but also the sum of $200,000 to endow "The John Shaw Billings Memorial Fund." Proceeds from this fund have been used, in part, to purchase special works that the Library could not otherwise afford. They include the first editions of Euclid's *Elements* (1482) and Newton's *Principia* (1687); the first edition of Thomas Geminus's reprinting, using copperplate engraving, of Vesalius's *Fabrica* (1545); and the second edition, but first French translation, of Charles Estienne's *La dissection des parties du corps humain* (1546). As the fund bookplate states, John Shaw Billings will long be remembered by The New York Public Library for his "foresight, energy, and administrative ability [that] made effective its far-reaching influence."

The Spencer Collection of illustrated books in fine bindings came to the Library in 1913 after the death of William Augustus Spencer on the *Titanic*, on April 14, 1912. The original collection has grown considerably, thanks to a fund that came with the books and carried the directive to purchase "the finest illustrated books in fine bindings that can be procured of any country and in any language, and to be bound in handsome bindings, representing the work of the most noted bookbinders of all countries, thus constituting a collection representative of the arts of illustration and bookbinding." Among the Spencer Collection books in *Seeing Is Believing* are the first editions of Joannes de Ketham's *Fasciculo di medicina* (1493), Albrecht Dürer's *Unterweysung der Messung* (1525), Otto Brunfels's *Herbarum vivae eicones* (1530), and Leonhart Fuchs's *De historia stirpium* (1542); and one of the dozen known copies of Anna Atkins's *British Algae* (1843–53), the work of the first woman photographer.

Over the course of the past century, the Library's scientific collections have continued to grow. In 1934, the Library's science and technology collections were augmented by the acquisition of the library of William Barclay Parsons, the engineer for the New York City subway system. The gift of Mrs. Parsons, the collection is devoted to engineering and transportation, and includes a great many rare books, such as Niccolò Tartaglia's *Nova scientia inventa* (1537); Vannoccio Biringuccio's *De la pirotechnia* (1540); the first printed edition of the works of Archimedes (1544); Agostino Ramelli's *Le diverse et artificiose machine* (1588) with the bookplate of Nicolai Joseph Foucault; and Domenico Fontana's *Della trasportatione dell'Obelisco Vaticano* (1590).

Most recently and notably, the Library received in 1995 the gift of the Wheeler Collection of Electricity and Magnetism. This gift from the United Engineering Trustees consisted of the library formed by Josiah Latimer Clark, which represented one of the most complete collections of books and periodicals on the subject of electricity assembled in the nineteenth century. As Clark observed the year before his death: "I have been collecting everything I can find in all languages for forty-

Francesco Bravo's *Opera medicinalia*, printed in Mexico in 1570, represents two milestones in the history of printing: it is the first medical book and the first illustrated book on any subject to have been printed in the New World. Of the three woodcut illustrations, the two shown here are of the sassafras plant. The woodcut on the right (*Smilax aspera*) is based on the one in Mattioli's *Commentaries on Dioscorides* (1554), a copy of which was in the library of the Franciscan convent of San Fernando in Mexico City in Bravo's time. The woodcut on the left, of the Mexican sassafras, was drawn from life in the New World. The third woodcut in the book reproduces a partial view of the vein system as described by Vesalius, providing some insight into the influence of Vesalius's work.

seven years. In that long time (during which I kept a skilled librarian) I succeeded in getting all English books both old and new. I also got a very large quantity of all foreign works, especially the rarer and older ones. In the line of pamphlets connected with early telegraphy my collection is quite unique, and comprehends 125 volumes. Although I still search catalogues, I rarely find anything that I have not got."

The collection was housed for many years in Clark's offices in Westminster, but it was his wish that his library should eventually go to the United States, since the library of his colleague, friend, and rival collector Sir Francis Ronalds was to stay in London. The move of the collection was accomplished through the good offices of Schuyler Skaats Wheeler, who purchased the Latimer Clark Library in 1901 and presented it to the American Institute of Electrical Engineers in New York City. In 1903, Andrew Carnegie contributed funds to house, catalog, and add to the Latimer Clark Library, which became known as the Wheeler Gift. Both the Engineering Societies' Building, including two floors for the library, and the two-volume catalogue were completed by 1909.

As one of the stipulations in his gift of the Latimer Clark Library to the American Institute of Electrical Engineers, Mr. Wheeler required that the "Library remain in New York City and . . . be a reference library, free to all." In 1995, when the overseers of the collection, the United Engineering Trustees, decided to give up running a library, the Wheeler Gift came to The New York Public Library, where it is administered by the Rare Books Division.

Important Wheeler books in *Seeing Is Believing* include William Gilbert's monumental *De magnete* (1600); Otto von Guericke's *Experimenta nova (ut vocantur) Magdeburgica de vacuo spatio* (1672); Luigi Galvani's *De viribus electricitatis in motu musculari* (1791); and James Clerk Maxwell's *Treatise on Electricity and Magnetism* (1873).

These are but a few of the major holdings in science and medicine that can be found in The New York Public Library. As the exhibition checklist in this publication attests, *Seeing Is Believing* presents a window onto the riches to be found in the Library's four research centers.

Seeing Is Believing

700 Years of Scientific and Medical Illustration

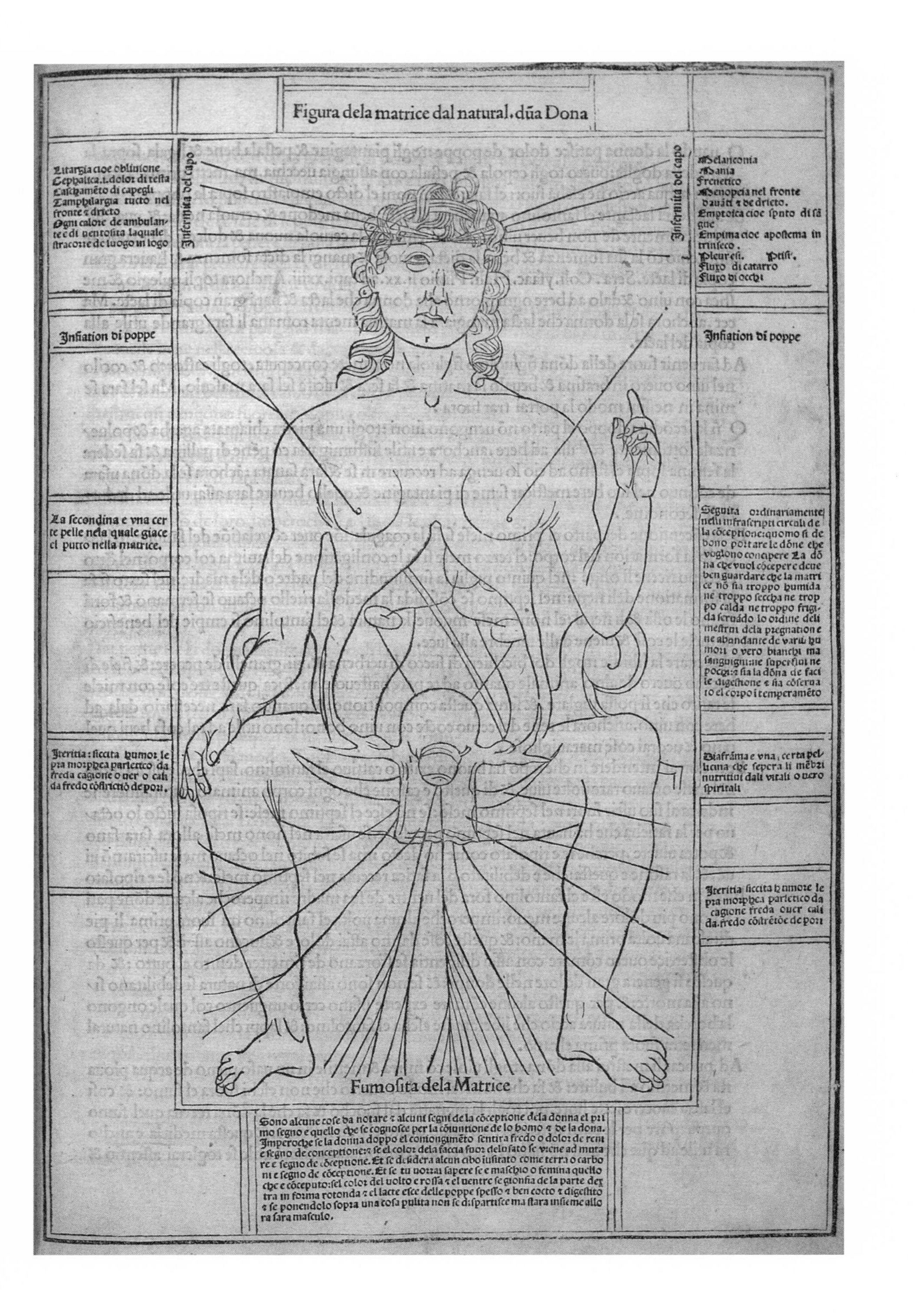

Figura dela matrice dal natural. dũa Dona
Litargia cioe oblinione
Cephalica. i. dolor di testa
Caschamẽto di capegli
Camphilargia tucto nel fronte ꝛ drieto
Ogni calore de ambulante e di uentosita laquale stracorre de luogo in logo
Infermita del capo
Infermita del capo
Melanconia
Mania
Frenetico
Menopeia nel fronte dauãti ꝛ de drieto.
Emptoica cioe spnto di sãgue
Empima cioe apostema intrinseco.
Pleuresi.
Ptisi.
Flugo di catarro
Flugo di occhi
Infiation di poppe
Infiation di poppe
La secondina e vna certe pelle nela quale giace el putto nella matrice.
Seguita ordinariamente nelli infrascripti circali de la cõceptione: quomo si debono portare le dõne che vogliono concipere La dõna che vuol cõceper e deue ben guardare che la matrice nõ sia troppo humida ne troppo secca ne troppo calda ne troppo frigida seruãdo lo ordine deli mestrui dela pregnatione e ne abandante de varii humori o vero bianchi ma sanguigni: ne superflui ne pochi: e sia la dõna de facile digestione e sia cõseruato el corpo i temperamẽto
Iteritia: siccita humor lepra morphea parletico da freda cagione o uer o calida fredo cõstrictiõ de pori.
Diafrãma e vna, certa pellicina che sepera li mẽbri nutritiui dali vitali o uero spiritali
Iteritia siccita hnmore lepra morphea parletico da cagione freda ouer calida. fredo cõstrẽtõe de pori
Fumosita dela Matrice
Sono alcune cose da notare ꝛ alcuni segni de la cõceptione dela donna el primo segno e quello che se cognosce per la cõiunctione de lo homo ꝛ de la dona. Imperoche se la donna doppo el coniongimẽto sentira fredo o dolor de reni e segno de conceptione: ꝛ se el color dela faccia fuor delusato se viene ad mutare e segno de cõceptione. Et se desidera alcun cibo iusitato come terra o carboni e segno de cõceptione. Et se tu uorrai sapere se e maschio o femina quello che e cõceputo: sel color del uolto e rossa ꝛ el uentre se gionfia de la parte dextra in forma rotonda ꝛ el lacte esce delle poppe spesso ꝛ ben cocto ꝛ digesito ꝛ se ponendolo sopra una cosa pulita non se dispartisce ma stara insieme allora sara masculo.

The Medieval Worldview

THE MEDIEVAL UNIVERSE was a contained place; its dimensions were known and understood. The Earth, an imperfect sphere, sat at its center, stationary but changeable, surrounded by crystalline spheres whose circular movements created a celestial harmony. The only truly immutable body in the medieval Christian universe was God, the source of all change on Earth and of movement in the heavens. Man, created in God's image, was the measure of all things terrestrial, that is, those things below the sphere of the moon.

The traditional images of medieval science and medicine range from the concentric circles of the known universe in Sacro Bosco's *De sphaera* (ca. 1275), to the seated anatomical figure of a woman in a fifteenth-century woodcut from Joannes de Ketham's *Fasciculo di medicina* (1493), to the diagrams and figures of Euclidean geometry; these illustrations had accompanied texts from the time of antiquity.

The *Fasciculus medicinae* (1491) was the first printed illustrated medical book; subsequent editions appeared in Italian (1493; shown here) and again in Latin (1495). Medical historians conjecture that Joannes de Ketham, or Johann of Kircheim, collected the text and anatomical figures, created about a hundred years earlier, for use in his medical lectures; the collection was simply given his name. The woodcuts, attributed to the school of Gentile Bellini, are often cited as among the greatest examples of fifteenth-century Venetian book illustration, and offer Renaissance renderings of stock medieval anatomical figures: here, a seated woman with an open thoracic cavity. This was the first depiction of the female viscera in a printed book.

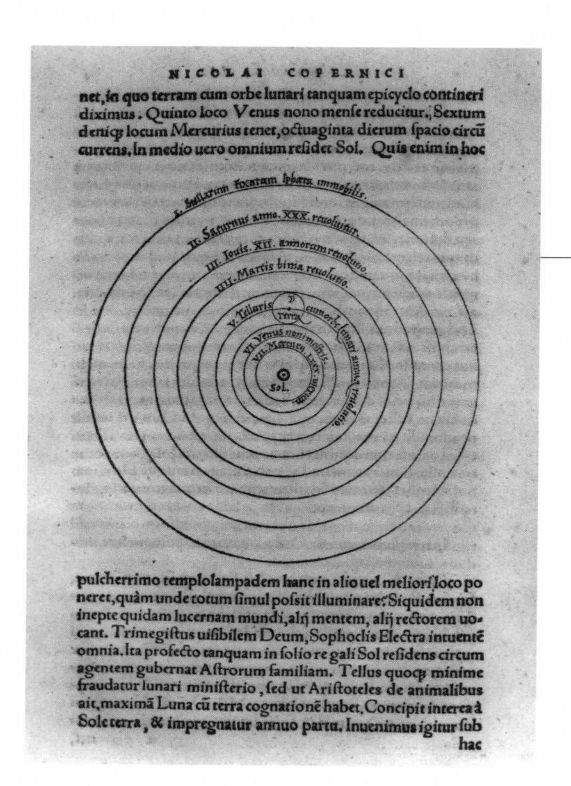

NICOLAI COPERNICI

net, in quo terram cum orbe lunari tanquam epicyclo contineri diximus. Quinto loco Venus nono mense reducitur. Sextum denicp locum Mercurius tenet, octuaginta dierum spacio circũ currens. In medio uero omnium residet Sol. Quis enim in hoc

pulcherrimo templo lampadem hanc in alio uel meliori loco poneret, quàm unde totum simul possit illuminare? Siquidem non inepte quidam lucernam mundi, alij mentem, alij rectorem uocant. Trimegistus uisibilem Deum, Sophoclis Electra intuentẽ omnia. Ita profecto tanquam in solio regali Sol residens circum agentem gubernat Astrorum familiam. Tellus quocp minime fraudatur lunari ministerio, sed ut Aristoteles de animalibus ait, maximã Luna cũ terra cognationẽ habet. Concipit interea à Sole terra, & impregnatur annuo partu. Inuenimus igitur sub hac

The most justly famed illustration of Western science, this image in the first printed edition of *De revolutionibus orbium coelestium* (1543) shows the Sun at the center of the universe. The work did not see print until the last year of Copernicus's life. He had good reason to fear the consequences of publication; *De revolutionibus* remained on the Vatican's Index, a list of prohibited books, until 1835.

Some of the illustrations and texts were rescued from oblivion by the Arabs and were augmented with the commentaries and observations of such major Arabic scientists as Avicenna and Rhazes. But by and large, the traditional images of the constellations and anatomical figures were retained and copied until the end of the seventeenth century in Persia and other areas of the Islamic world. It was the translation work of ninth-century Arabs and Syriac Jews, in particular, that succeeded in preserving the "lost" classical works, such as those of Aristotle. The influx of these manuscripts into Europe, occasioned by the retreat of the Moors in Spain and the Crusades, brought European Christian scholars to recognize the need to integrate these classical texts into their own cosmology and subsequently into their science and medicine. But by 1543, the year of the publication of Vesalius's *De humani corporis fabrica* and Copernicus's *De revolutionibus orbium coelestium*, classical science and medicine were no longer adequate to serve the needs of the European intellectual community.

In astronomy, for example, was it conceivable that an omnipotent Christian God would create a universe that could be explained

"Neither Renaissance nor Reformation was the movement that produced the really great revolution from the medieval to the modern world; that was effected by the gradual development of science."

John Herman Randall, Jr. (1899–1980)

only by so difficult and convoluted a system as that of Ptolemy? To Copernicus, the simpler and mathematically more elegant explanation that put the Sun at the center of the universe made far more sense. Recognizing the revolutionary nature of his theory, he tested the waters by circulating it in manuscript form as the *Commentariolus*; most of the great sixteenth-century scientists, including Tycho Brahe and Johannes Kepler, read it first in that form. His completed work, *De revolutionibus orbium coelestium,* was finally published at the end of his life, in 1543. But it was not until the end of the seventeenth century that the heliocentric universe became indisputable fact.

Vesalius, on the other hand, revolutionized medicine by demonstrating in his teaching at the University of Padua and through his major work, *De humani corporis fabrica* (1543), that human anatomy should be learned directly from the dissection of human bodies. He also insisted that physicians do their own dissecting and not leave the work to assistants. But cadavers, even those of executed criminals, were not readily available, and most people were averse to the idea of allowing their friends and relatives to be cut up in the service of medicine or art. Luckily for Vesalius, Marcantonio Contarini, the judge of the Paduan criminal court, directed that the bodies of executed criminals be made available to him for dissection. In this way, Vesalius was able to work toward his major ambition to revolutionize the teaching of anatomy and overthrow the reliance on the accepted authority of Galen (131–201 C.E.), probably the most famous classical physician after Hippocrates and the most influential, whose anatomy was based on animal dissections.

In botany, too, things were changing. The primary classical authority in the field was Dioscorides, a Greek who lived during the first century C.E., during the reigns of Claudius and Nero, and who may have served as a physician in the Roman army. He was the first to write on medical botany, and his book circulated in manuscript form beginning around 78 C.E. From the sixth century on, many of the manuscripts were illustrated, providing some of the most important early depictions of plants. However, through continued copying, many of these images became corrupted, a process that only became worse after the first printed edition appeared in 1478.

Otto Brunfels's *Herbarium vivae icones* initiated a new standard for herbal illustration with its publication in 1530. Hans Weiditz, the artist, drew the plants from nature in a marked departure from the conventional images characteristic of previous herbals. But Brunfels's text is limited and hackneyed, having been based on second- or third-rate studies of local plants that followed the lead of the classical authorities, primarily Dioscorides. This inspired Leonhart Fuchs to publish his great herbal in 1542, going beyond Brunfels to show not only the largest number of plants known to be useful as drugs and herbs, but also the characteristics and habitats of the plants themselves.

Most of the woodblocks used for the 1543 and 1555 printings of Vesalius's *De humani corporis fabrica* and for the *Epitome* survived into the twentieth century. They were found in 1932 in the University of Munich Library by Dr. Charles W. Lester, President of The New York Academy of Medicine, who arranged for them to be printed by the Bremer Presse in 1935. This limited edition includes impressions of all the known blocks, including this frontal view of one of the fifteen flayed muscle figures, which illustrate successive muscle layers down to the skeleton; when placed side by side, the backgrounds form a continuous panoramic landscape. Unfortunately, all these blocks were subsequently destroyed by the Allied bombing of Munich during World War II.

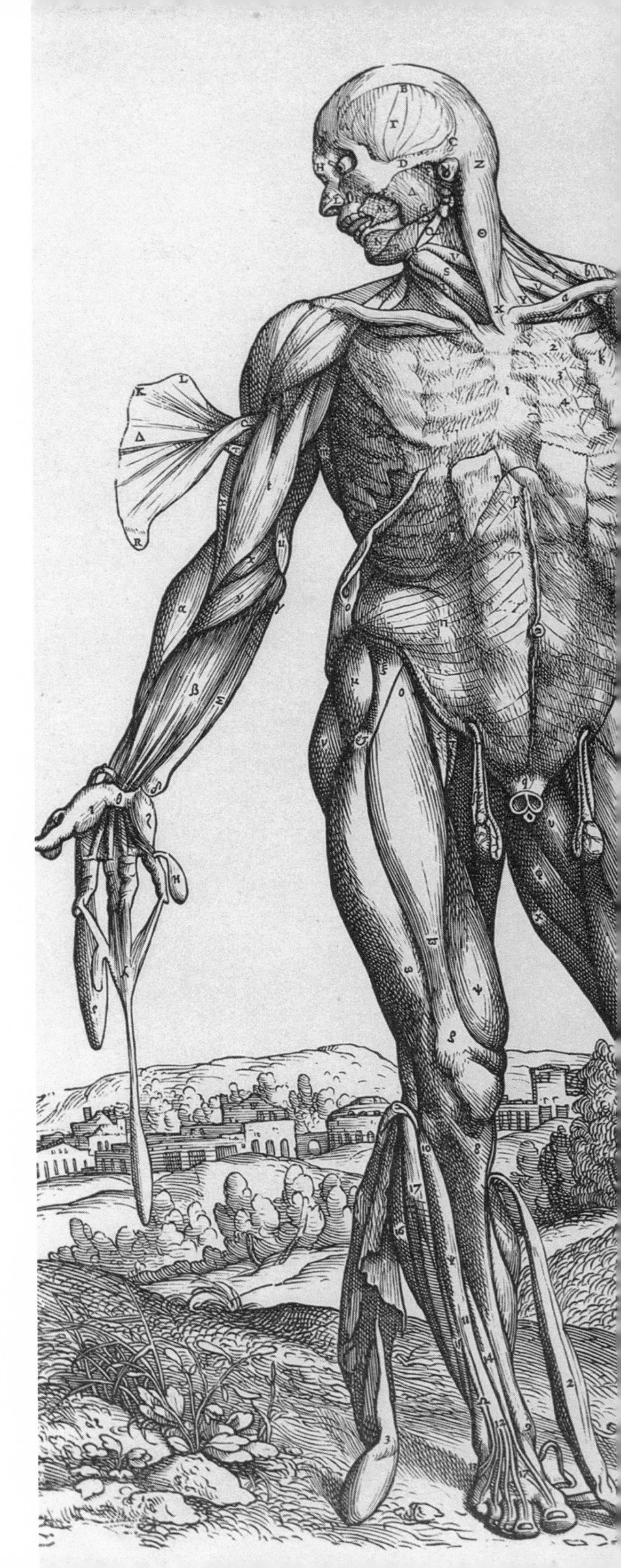

Although Fuchs, a medical professor at the University of Tübingen, did not recognize the relationship between plants or of plants to other living things, he did realize that certain species were found only in specific regions, including the fact that certain plants described by Dioscorides did not grow north of the Alps, while other plants in the north were unknown to the ancient authorities. Published the year before Copernicus's *De revolutionibus* and Vesalius's *Fabrica*, Fuchs's work was revolutionary in that it began to dispel the authority of Dioscorides. Not as controversial as the works that overturned the authority of Ptolemy and Galen, the book went through thirty-five editions during the author's lifetime.

It was discoveries and arguments such as these that set the stage for the scientific revolution of the seventeenth century.

PHILOSOPHORVM ΣΟΦΩΤΑΤΩ ÆSCVLAPIO SVO.

Viuere cui vires, & robora ſana dediſti,
Scribere ni vellem, næ robore durior eſſem.
Ergo mihi (quæ priuato pertingere nulli
CASE datur) tecum ſatis & ſatis Aſtra tueri eſt,

Astronomy

THE SIXTEENTH CENTURY ended and the seventeenth century began with the burning of Giordano Bruno as a heretic by the Inquisition in 1600. He had proposed the possibility of a plurality of inhabited worlds and an infinite universe, concepts that were dangerously at odds with traditional medieval cosmology. But it was almost impossible to avoid theorizing about the nature of the stars and planets, because astronomical phenomena simply did not fit the existing theories. And the dichotomy between accepted theory and observation became more and more obvious as methods of observation and calculation became more sophisticated.

Until Galileo, no one had used a telescope to look at the heavens; the work of Tycho Brahe and Johannes Kepler was based entirely on observations with the naked eye. Galileo was the first to use an astronomical telescope, and although his discoveries and conclusions, first

John Case's *Sphaera civitates* first appeared in 1588, shortly after the English victory over the Armada of Spain, and was reprinted the following year in Frankfurt. This illustration from that edition celebrates Elizabeth I by showing the civic universe in the guise of the old universe scheme. Immovable justice appears at the center in place of the immovable Earth. The seven spheres show abundance as the moon, eloquence as Mercury, clemency as Venus, religion as the Sun, fortitude as Mars, prudence as Jupiter, and majesty as Saturn. The sphere of the fixed stars contains the Star Chamber, Nobles, Lords, and Councilors. Elizabeth herself appears, in place of the Deity, as the Prime Mover of the spheres of the state.

published in 1610 in the *Sidereus nuncius*, did not cost him his life, they did lead to his confinement in a kind of house arrest until his death. Galileo's observations revealed the following: the moon was not a perfect sphere, it had mountains and valleys like the Earth; Jupiter had at least three satellites, whose existence disproved the assertion that the heavenly bodies revolved exclusively around the Earth; and there were observable phases of the planet Venus, a phenomenon that could be explained only if Venus moved around the Sun, not the Earth. *Sidereus nuncius* thus became a milestone astronomical text; Galileo's drawings of the surface of the moon were repeated (first in woodcuts, then in engravings) in all the major astronomical works until the eighteenth century, and the telescope became standard operating equipment for astronomers.

The problem of motion in the heavens and on Earth preoccupied scientific thinkers and provided the intellectual battleground for the war of the world systems that was waged throughout the seventeenth century. Initiated by Copernicus's *De revolutionibus,* the battle waxed and waned in the arena of astronomy and physics. Besides providing reports of telescopic observation that offered indisputable proof of celestial phenomena, texts such as *Sidereus nuncius* also seriously challenged the peculiar nature of the Aristotelian, i.e., medieval, universe, in which all motion had to be accompanied by a mover or the Prime Mover, God.

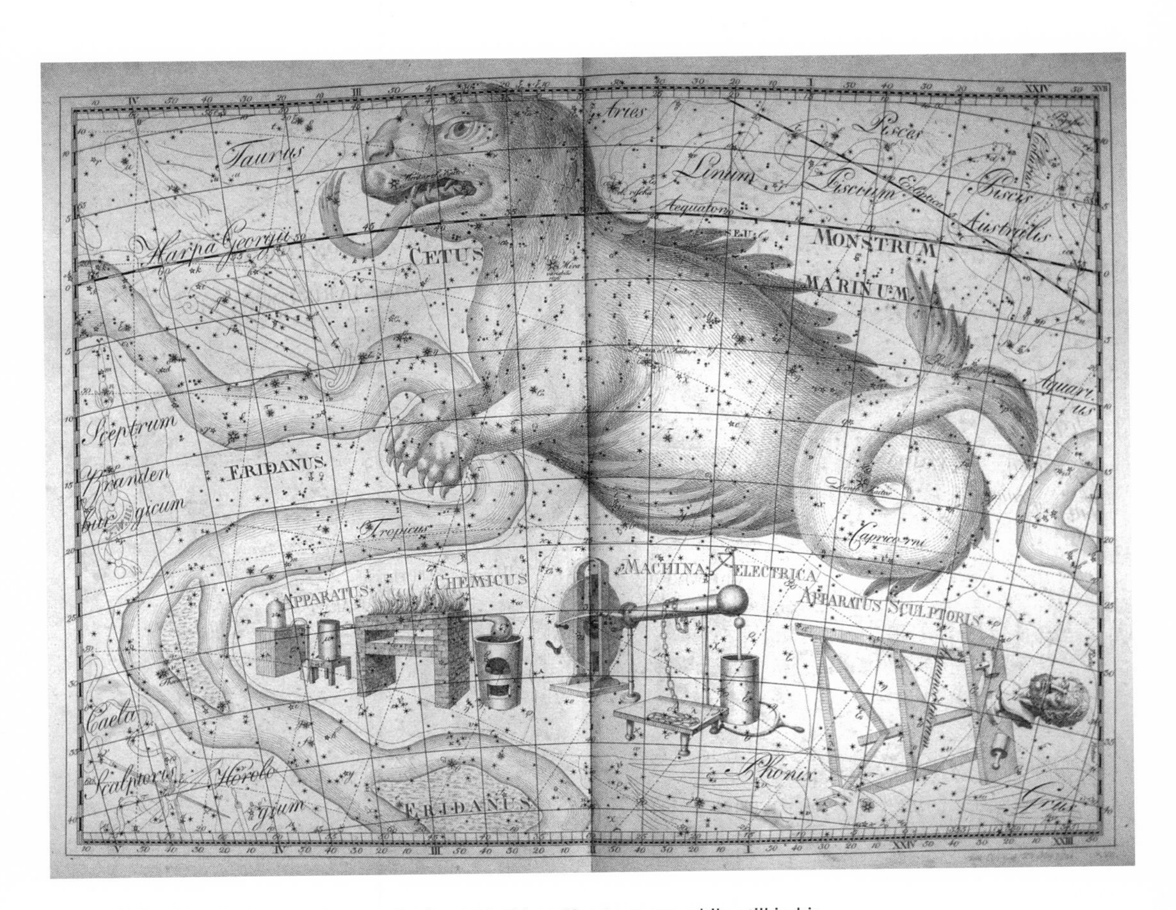

A gifted mathematician, Johann Elert Bode taught himself astronomy while still in his teens. From 1774 until his death in 1826, he produced a series of well-received astronomical almanacs for the observatory of the Berlin Academy, where he was appointed director and royal astronomer in 1786. His *Uranographia* (1801), the most comprehensive sky atlas that had yet appeared, listed over 17,000 stars and 2,500 nebulae, and was the first to include the nebulae, star clusters, and double stars. This image shows chemical and electrical apparatus, as well as a new rendering of the constellation figures Cetus, depicted here not as a whale but as a sea monster.

"When the heavens were a little blue arch, stuck with stars, methought the universe was too straight and close: I was almost stifled for want of air: but now it is enlargd in height and breadth, and a thousand vortices taken in. I begin to breathe with more freedom, and I think the universe to be incomparably more magnificent than it was before."

Bernard Le Bovier (Sieur de) Fontenelle (1657–1757)

It was not until the end of the seventeenth century that the heliocentric universe became indisputable fact. Tycho Brahe had tried an unsuccessful compromise of sorts between the Ptolemaic and the Copernican systems, while Kepler hedged his conclusions, hiding them in the mathematics of his laws of planetary motion: (1) that the orbits of the planets are ellipses with the Sun as their focus, (2) that the line joining a planet to the Sun sweeps out equal areas in equal times, and (3) that squares of the periodic times are proportional to the cubes of the mean distances from the Sun. Although he became a strong advocate of the Copernican system and almost lost his life as a result, Galileo was unable to provide the definitive mathematical proofs to support his observations; that was left for Newton.

Once the medieval cosmos had been definitively overthrown and the problem of celestial motion solved, the telescope provided an enhanced and more accurate means of cataloging the number and magnitude of the stars, plotting the motion of the planets, and observing celestial phenomena. Celestial atlases, which actually date from about the middle of the sixteenth century, mapped the positions of the stars and generally pictured their various magnitudes by utilizing larger and smaller engraved figures. A grid overlay was introduced early on to allow for more accurate mapping, while the constellations that served as reference points were rendered in the traditional images of classical mythological figures.

By 1801, the year in which Johann Bode's *Uranographia* was published, Bode could provide the positions of about 17,000 stars and 2,500 nebulae in addition to plotting the motion of a new planet, Uranus, discovered by William Herschel in 1781 and given its name by Bode. By 1860, Angelo Secchi, the Director of the Roman Observatory, had photographed a solar eclipse, the Sun's corona, and Sun spots. His photographs appeared, reproduced by chromolithography, in the first journal on spectroscopy, *Memorie della Società degli Spettroscopisti Italiani*, edited by Pietro Tacchini; Secchi was a contributing editor. By the end of the nineteenth century, observatories in Europe and the United States routinely photographed various celestial phenomena and published those observations.

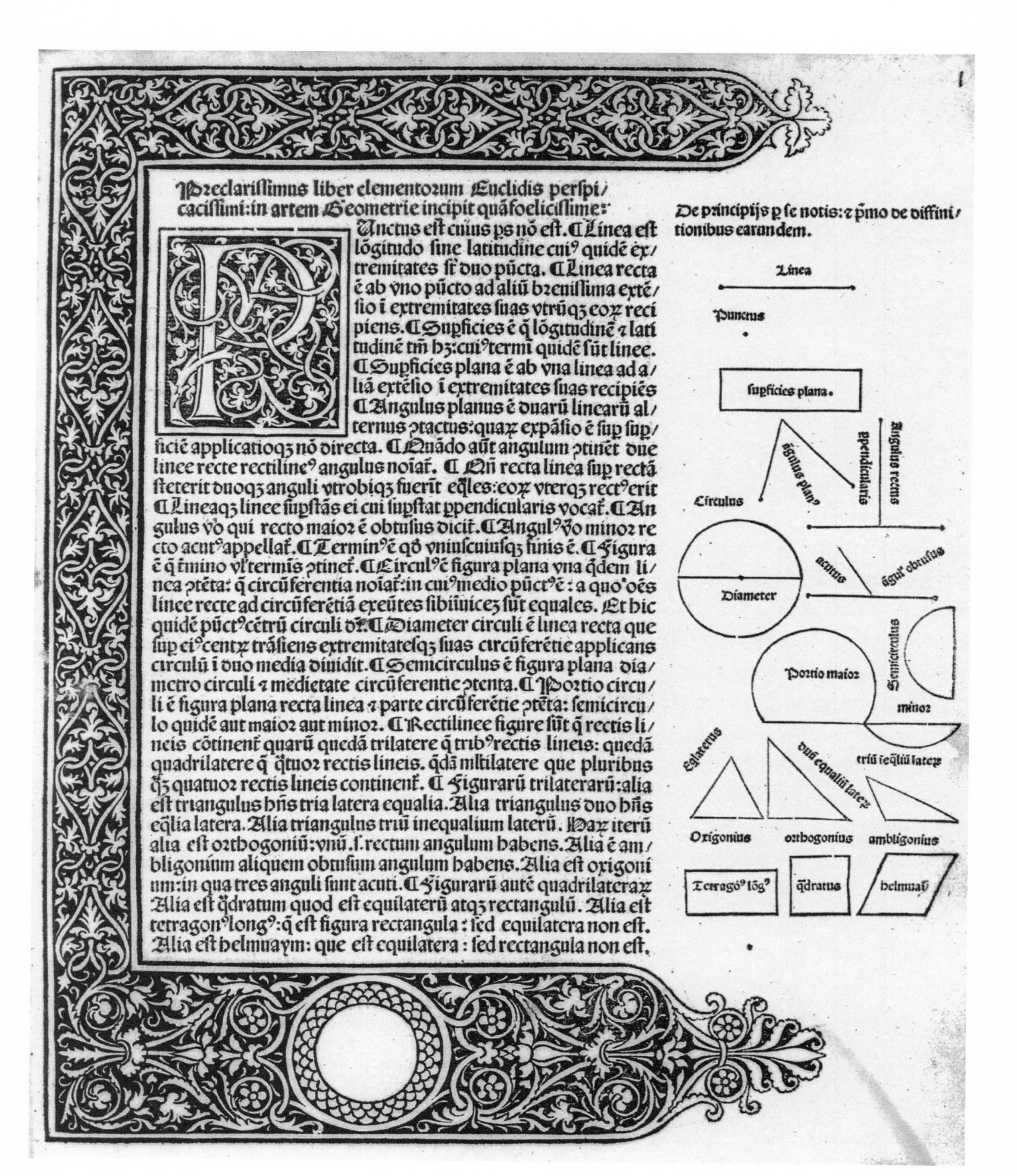

Preclarissimus liber elementorum Euclidis perspi/
cacissimi: in artem Geometrie incipit quã foelicissime:

Punctus est cuius ps nõ est. ¶ Linea est lõgitudo sine latitudine cuiꝰ quidẽ ex/tremitates sr̃ duo pũcta. ¶ Linea recta ẽ ab vno pũcto ad aliũ breuissima extẽ/sio ĩ extremitates suas vtrũqȝ eoꝝ reci/piens. ¶ Supficies ẽ q̃ lõgitudinẽ ⁊ lati/tudinẽ tm̃ hȝ: cuiꝰ termĩ quidẽ sũt linee. ¶ Supficies plana ẽ ab vna linea ad a/liã extẽsio ĩ extremitates suas recipiẽs ¶ Angulus planus ẽ duarũ linearũ al/ternus ꝯtactus: quaꝝ expãsio ẽ sup sup/ficiẽ applicatioqȝ nõ directa. ¶ Quãdo aũt angulum ꝯtinẽt due linee recte rectilineꝰ angulus noiat̃. ¶ Qñ recta linea sup rectã steterit duoqȝ anguli vtrobiqȝ fuerĩt eq̃les: eoꝝ vterqȝ rectꝰ erit ¶ Lineaqȝ linee supstãs ei cui supstat ppendicularis vocat̃. ¶ Angulus võ qui recto maior ẽ obtusus dicit̃. ¶ Angulꝰ võ minor recto acutꝰ appellat̃. ¶ Terminꝰ ẽ qd̃ vniuscuiusqȝ finis ẽ. ¶ Figura ẽ q̃ t̃mino vl̃ termĩs ꝯtinet̃. ¶ Circulꝰ ẽ figura plana vna q̃dem li/nea ꝯtẽta: q̃ circũferentia noiat̃: in cuiꝰ medio pũctꝰ ẽ: a quo oẽs linee recte ad circũferẽtiã exeũtes sibiĩuicẽ sũt equales. Et hic quidẽ pũctꝰ cẽtrũ circuli dr̃. ¶ Diameter circuli ẽ linea recta que sup eiꝰ centꝝ trãsiens extremitatesqȝ suas circũferẽtie applicans circulũ ĩ duo media diuidit. ¶ Semicirculus ẽ figura plana dia/metro circuli ⁊ medietate circũferentie ꝯtenta. ¶ Portio circu/li ẽ figura plana recta linea ⁊ parte circũferẽtie ꝯtẽta: semicircu/lo quidẽ aut maior aut minor. ¶ Rectilinee figure sũt q̃ rectis li/neis cõtinent̃ quarũ quedã trilatere q̃ tribꝰ rectis lineis: quedã quadrilatere q̃ q̃tuor rectis lineis. q̃dã m̃ltilatere que pluribus q̃ȝ quatuor rectis lineis continent̃. ¶ Figurarũ trilaterarũ: alia est triangulus hñs tria latera equalia. Alia triangulus duo hñs eq̃lia latera. Alia triangulus triũ inequalium laterũ. Haꝝ iterũ alia est orthogoniũ: vnũ .s. rectum angulum habens. Alia ẽ am/bligonium aliquem obtusum angulum habens. Alia est oxigoni/um: in qua tres anguli sunt acuti. ¶ Figurarũ autẽ quadrilateraꝝ Alia est q̃dratum quod est equilaterũ atqȝ rectangulũ. Alia est tetragonꝰ longꝰ: q̃ est figura rectangula: sed equilatera non est. Alia est helmuaym: que est equilatera: sed rectangula non est.

De principijs p se notis: ⁊ pmo de diffini/tionibus earundem.

Mathematics

THE MEDIEVAL SCHOLASTICS regarded mathematics as the limit of human reason, the highest point to which the rational human mind could aspire; only faith surpassed it in the hierarchy of intellectual disciplines. And even though mathematics could be successfully applied to the mundane needs of everyday life, mathematicians regarded such work as the domain of lesser intellects. This particular prejudice began with the Greeks – Archimedes, for example, was apologetic about the fact that his work had a practical application – and continued until well into the sixteenth century.

The pursuit of improved methods of calculating and measuring marked a departure from this view. As post-medieval thinkers became more and more preoccupied with investigations into the nature of physical phenomena, they became increasingly dependent upon mathematics for accurate description and prediction. Although pure mathematics continued to be an intellectual pursuit, and resulted in

---

The first of over a thousand printed editions of Euclid's *Elements*, this 1482 edition is the oldest scientific textbook still in use, and one of the most beautiful. The text is that of Campanus of Novara (d. 1296), who used a translation into Latin made early in the twelfth century by Adelard of Bath from an Arabic version of the Greek text. The printer, Erhard Ratdolt, included the definitions and diagrams of the geometric elements here on the first page of text. This book, with its many elegant white-on-black woodcut initial letters and diagrams placed in the margins to illustrate a particular portion of text, set the style for a great many of the scientific works that followed it.

significant advances, the scientific revolution of the seventeenth century established mathematics also as the language in which the results of the new observations and experiments in the physical sciences were expressed. Descartes and Newton used mathematics to describe and define space and movement in space, while others sought practical applications for mathematics in ballistics, engineering, and medicine.

By the seventeenth century, arithmetic and algebra had developed a notation system that is easily recognized today. All the major classical mathematical treatises had been printed, annotated, and improved upon, and aids to calculation such as logarithms had been developed. In its method of graphic design and illustration, as well as in its method of presentation of the text, the first edition of Euclid's *Elements* (1482) became the paradigm for mathematical works: in his dedication to the volume, the printer, Erhard Ratdolt, attributed the dearth of printed mathematical works to that time to the problems involved in printing the diagrams. While he did not explain his own method, of which he was justifiably proud, it involved the use of type-metal rules that could be printed at the same time as the text, along with woodcut diagrams. Both Descartes's *Principia philosophiae* (1644) and Newton's *Philosophiae naturalis principia mathematica* (1687) were written in the form of the *Elements*, and Euclidean geometry continued to define space until well into the nineteenth century. Only late in that century was space defined in non-Euclidean terms: parallel lines meet in curved space.

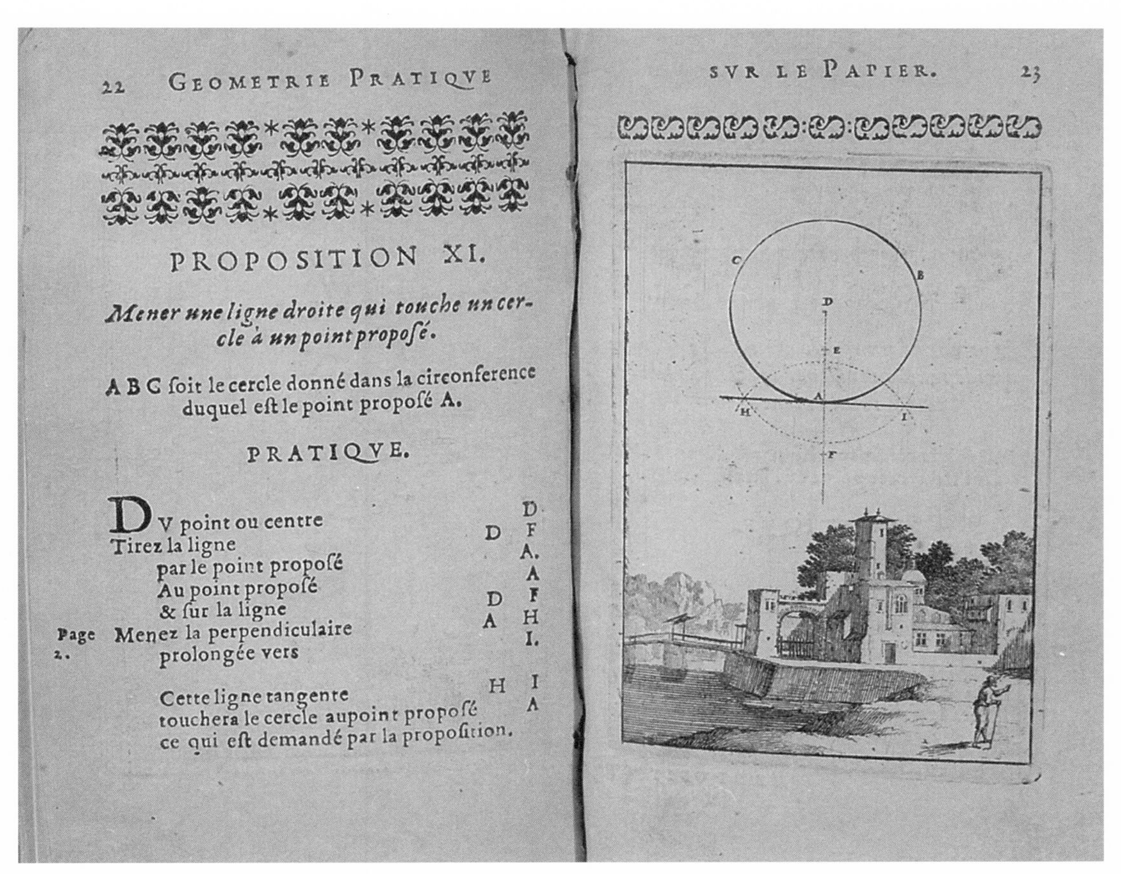

22 GEOMETRIE PRATIQVE

PROPOSITION XI.

Mener une ligne droite qui touche un cercle à un point propoſé.

A B C ſoit le cercle donné dans la circonference duquel eſt le point propoſé A.

PRATIQVE.

DV point ou centre — D
Tirez la ligne — D F
par le point propoſé — A.
Au point propoſé — A
& ſur la ligne — D F
Menez la perpendiculaire — A H
prolongée vers — I.

Page 2.

Cette ligne tangente — H I
touchera le cercle au point propoſé — A
ce qui eſt demandé par la propoſition.

SVR LE PAPIER. 23

An engraver, geographical engineer, and dabbler in physics, Sébastien LeClerc was instrumental in creating a style of vignette that became very popular in French books of the eighteenth century. Certainly his smaller illustrations, such as those in *Practique de la geometrie sur le papier et sur le terrain* (1669), a practical treatise on geometry and perspective meant for the use of artists and architects, demonstrate a grace and elegance of design tempered by a sense of humor. This popular book went through a number of editions and translations: there were editions in English, Dutch, Latin, and even Russian. The geometric figures float above exotic, often inhabited landscapes; this illustration shows *Proposition XI*, or, on drawing a tangent through a point on the circumference of a circle.

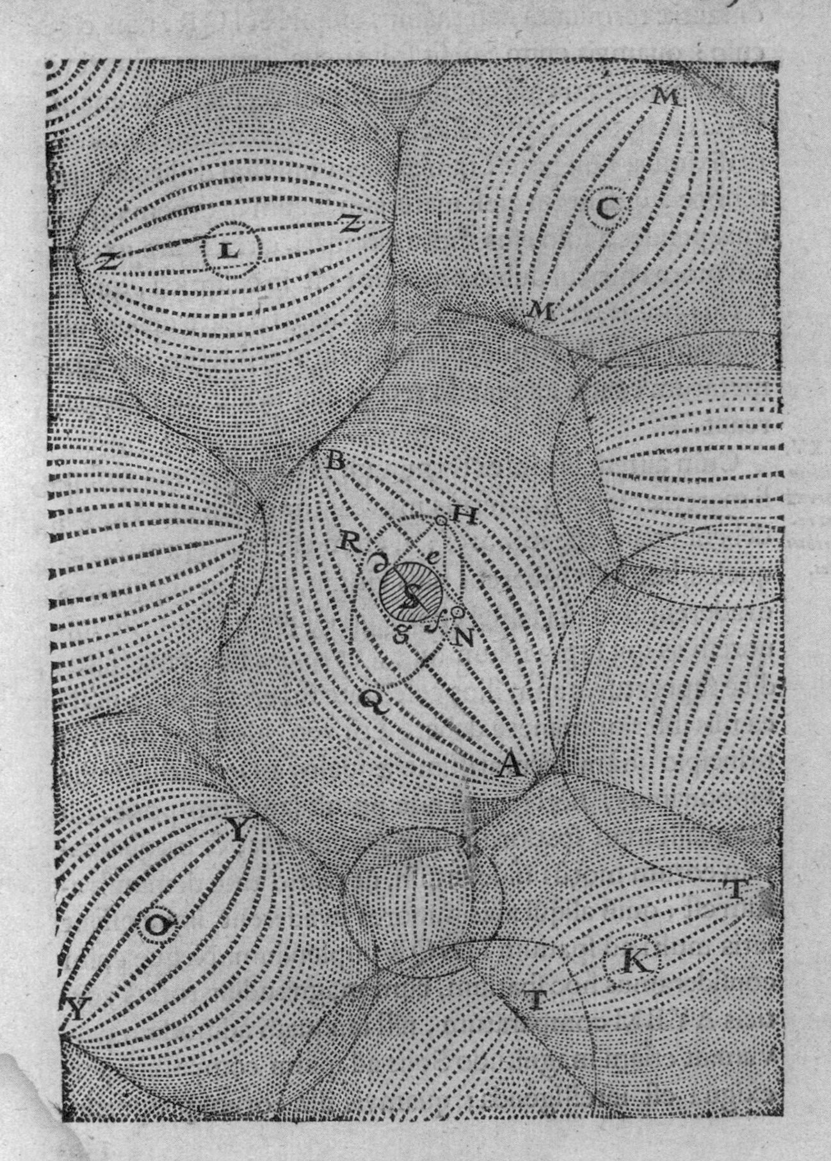
M
C
Z
L
Z
M
B
H
R
e
S
N
g
Q
A
Y
O
Y
T
K
T

Physics

WHAT IS INCLUDED in physics has changed radically over the centuries. To the Greeks, and throughout the Middle Ages, physics was considered the study of all natural phenomena, including chemistry, astronomy, meteorology, and biology as well as what is now called physics, but excluding mathematics, which was considered to be a superior intellectual endeavor.

From the time of the Greeks, the problems of motion and light had occupied scientific thinkers. Classical and medieval cosmology had approached light and movement philosophically. But in the Renaissance and the seventeenth century, when medieval cosmology failed to adequately account for observed physical phenomena, motion and light were approached through work in the fields of mechanics and optics, in which the problems were defined and eventually solved experimentally and mathematically.

Like everyone else interested in physics in the seventeenth century, Descartes set out to solve the problem of motion. Observation and experimentation had shown that classical physics was inadequate to explain physical phenomena, particularly movement in the heavens. If space was empty, who or what was moving the planets? Descartes could not accept the possibility of movement in a void. To fill space and then account for the motion of the planets, he proposed an aethereal substance that he called *plenum*, and vortices (whirlwinds) of plenum to provide the impetus and gravitational effects that forced the planets into elliptical orbits. This illustration from his *Principia philosophiae* (1644) shows Cartesian space, with depressions marking the vortices.

"There are more things in heaven and earth, Horatio,
Than are dreamt of in your philosophy."

SHAKESPEARE, *HAMLET*, ACT I, SCENE V

In his magnum opus, *De magnete* (1600), the first significant body of scientific work produced and published in England, William Gilbert recorded every available fact about magnetism, attempting to test every statement empirically. He described the Earth as a giant magnet, and gave an account of experiments in which he placed a sphere, the "Terrella" or "Little Earth" (one of his favorite experimental devices), in a magnetic field, where it revolved. The accounts of these experiments were crucial to Galileo's speculations on the nature of celestial and terrestrial mechanics, as well as to the work of Kepler, Boyle, and Descartes.

René Descartes set out to solve the problems of motion and light in two works: the *Principia philosophiae* (1644), and *La dioptrique*, published in his *Discours de la méthode pour bien conduire sa raison* (1637). In the *Principia,* he presented a system of physics based on a purely mechanistic explanation of the universe. He actually came quite close to solving the problem of motion – his concept of the state of motion and the law of inertia would become central to Newton's work – but the flaw in his principles was the lack of mathematical rigor; his *Principia* was based on philosophical principles. Where Newton would promulgate the laws of motion based on his understanding of the nature of mass, Descartes founded his principle of inertia on the notion that what God had set in motion in the beginning could not be destroyed.

Although his *Discours*, like the *Principia*, relies on rational and non-empirical reasoning from first principles, in each of the three works

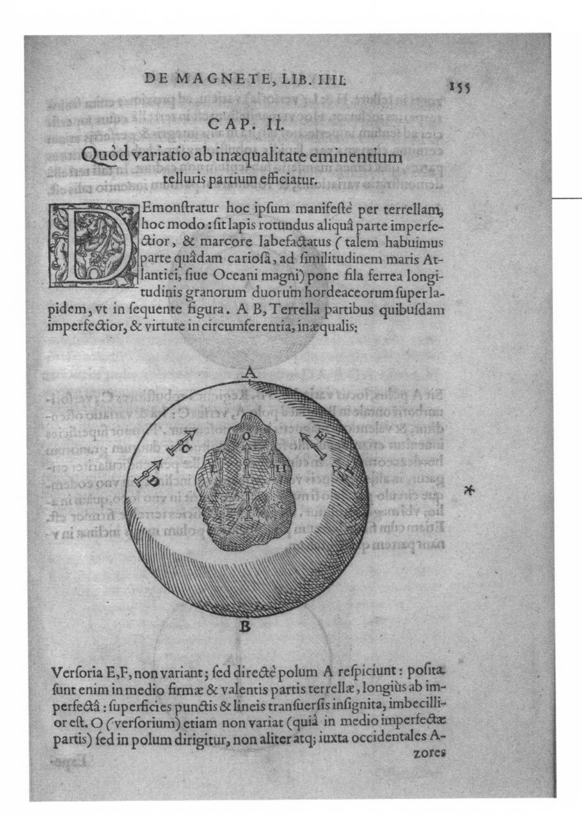

DE MAGNETE, LIB. IIII. 155

CAP. II.

Quòd variatio ab inæqualitate eminentium telluris partium efficiatur.

Demonſtratur hoc ipſum manifeſtè per terrellam, hoc modo: ſit lapis rotundus aliquâ parte imperfectior, & marcore labefactatus (talem habuimus parte quâdam carioſâ, ad ſimilitudinem maris Atlantici, ſiue Oceani magni) pone fila ferrea longitudinis granorum duorum hordeaceorum ſuper lapidem, vt in ſequente figura. A B, Terrella partibus quibuſdam imperfectior, & virtute in circumferentia, inæqualis;

Verſoria E,F, non variant; ſed directè polum A reſpiciunt: poſitæ ſunt enim in medio firmæ & valentis partis terrellæ, longiùs ab imperfectâ: ſuperficies punctis & lineis tranſuerſis inſignita, imbecillior eſt. O (verſorium) etiam non variat (quià in medio imperfectæ partis) ſed in polum dirigitur, non aliter atq; iuxta occidentales Azores

Before becoming royal physician to Queen Elizabeth I, William Gilbert had an established reputation as an expert on maritime diseases and an interest in the phenomenon of electric and magnetic attraction. Elizabeth I was also interested in magnetism, particularly in experiments involving the compass, because of her desire to improve English navigation and ensure English control of the seas. Their interests dovetailed nicely because the compass constituted the only practical application for magnetic theory at the time. Gilbert published the results of his experiments in *De magnete* (1600); this illustration is from Book IV, which focuses on experiments to account for magnetic variations due to differences in the Earth's elevation. He hypothesized that such variations accounted for the variation in compass readings for a particular location.

published with the *Discours* (*La géométrie, La dioptrique*, and *Les météores*), Descartes demonstrated a very modern reliance on experimentation, observation, and laborious mathematical calculation. In *La dioptrique*, for example, he attempted to demonstrate how light is perceived. Descartes patiently dissected the eyes of numerous animals and humans and combined those observations with his knowledge of mathematics and refraction to propose, for example, that the lens of the eye changes shape in order to accommodate distance and that light is perceived because of the specific movement of parts of the eye to accommodate it.

By the end of the seventeenth century, the known universe had expanded both macroscopically, with Galileo's invention of the telescope, and microscopically, with Robert Hooke's invention of the compound microscope. Man was no longer the measure of all things terrestrial, and God was set at some distance from His creation: the Sun had replaced the Earth at the center of the universe, and

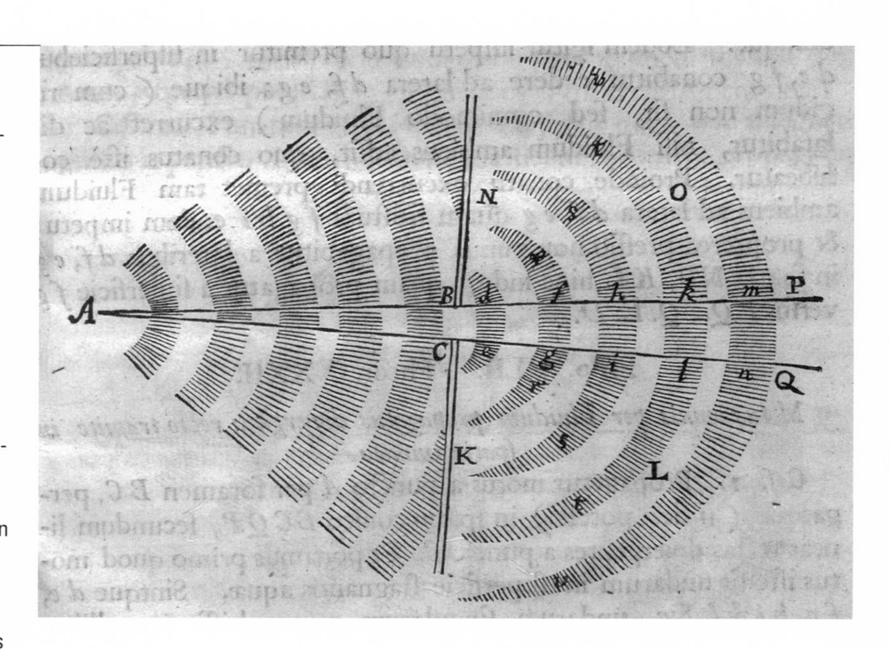

By the late seventeenth century, Kepler's laws of planetary motion notwithstanding, the problem of motion, terrestrial and celestial, had not been solved. Edmond Halley asked Newton to work on this problem, and the *Philosophiae naturalis principia mathematica* (1687) was the result. As the Royal Society was unable to fund its publication, Halley paid for the work himself and saw it through the press. Written entirely in Latin, and so difficult as to be considered incomprehensible by most, the *Principia* has the distinction of being the most influential, yet least-read book in the history of science. This illustration relates to the movement of a wave through an aperture, denoted by the letters B and C.

it had been determined that the heavenly bodies and the Earth were governed by the same physical principles. Furthermore, those physical principles or laws could be expressed mathematically.

The man whose work epitomized this revolution was Sir Isaac Newton. He not only developed a methodology based on experimentation and critical observation that made it possible to delineate physical phenomena mathematically, he also invented the mathematics that made it possible to accurately predict those phenomena. After the publication of Newton's *Principia* (1687) and other works, there was virtually no area of intellectual discourse, including the political, that did not make reference to him.

Although Newton had by then provided the laws of motion, in the eighteenth century light remained an important area of inquiry, as did electricity and magnetism, meteorology, hydrodynamics and hydrostatics, and mechanics. Electricity in particular captured the popular and scientific imagination of the late eighteenth and early

William Konrad Röntgen, a professor of physics and director and later rector of the Physical Institute of the Royal University of Würzburg, made his first announcement of the discovery of what he called X rays on December 28, 1895, to the editors of the Physical and Medical Society of Würzburg. The *Wiener Presse* carried the story on January 5, 1896, and the news broke around the world the following day. Shown here, from the *Illustrierte Zeitung* of August 22, 1896, is the first full-skeleton X-ray image to appear in the popular press.

nineteenth centuries. From Mary Shelley to Benjamin Franklin, in published fiction and in papers presented before the Royal Society in London, the American Philosophical Society in Philadelphia, and the Académie des Sciences in Paris, it seemed that almost everyone was experimenting with electricity, passing electric currents through anyone and anything and creating incredible machines to generate wonderful arcs of current, smoke, and static. All of this activity was chronicled in illustrations. Even Giovanni Aldini's experiments on electricity – which involved passing an electric current through the dead body of an executed criminal and the severed head of a cow – were accompanied by graphic illustrations (in Aldini's experiments, the cadaver moved and opened its eyes, as did the cow). His *Account of the Late Improvements in Galvanism* (1803) might have cost him his life in an earlier

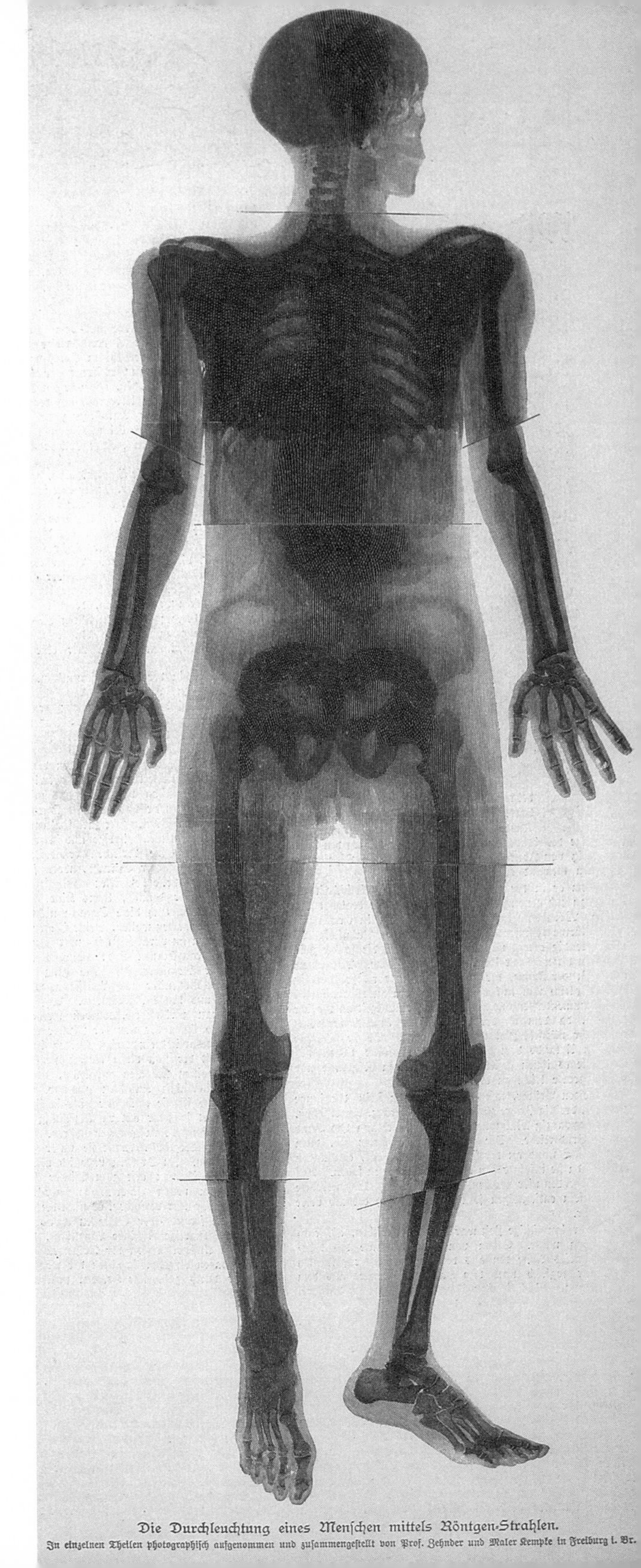

Die Durchleuchtung eines Menschen mittels Röntgen-Strahlen.
In einzelnen Theilen photographisch aufgenommen und zusammengestellt von Prof. Zehnder und Maler Kempke in Freiburg i. Br.

Della trasportatione dell'Obelisco Vaticano (1590), one of the finest of all engineering books, describes and illustrates, in thirty-eight plates, one of the most famous events in engineering history, Domenico Fontana's moving of the magnificent 361-ton Egyptian obelisk from its position behind the Vatican Sacristy to the piazza in front of St. Peter's Basilica in 1586. Fontana supervised the work himself, using trumpet calls and bells to synchronize the work of 900 men and seventy-four horses. Fontana's drawings, engraved by Natali Bonifazio da Sabenico, were sold as souvenirs during the procedure, which took four and a half months to complete.

time, but in the early nineteenth century simply inspired a classic story, *Frankenstein* (1818).

The nineteenth century was again a period of heightened inquiry into the problems of light and motion. The ability to redefine space mathematically through non-Euclidean geometry, and investigations into the nature of light by Michael Faraday, James Clerk Maxwell, Heinrich Hertz, and others, resulted in many new advances in physics. Some of these advances included the laws of conservation of energy, the wave theory of light, the discovery of radiation and the absorption of heat, the extension of radiation into the infrared and ultraviolet portions of the spectrum, spectroscopy, and the ability to penetrate

In 1879, when the Khedive of Egypt presented the 220-ton obelisk known as Cleopatra's Needle to the United States, Lieutenant-Commander H. H. Gorringe, a U.S. Navy engineer, volunteered to oversee its removal from Alexandria to New York City. *Egyptian Obelisks* (1882) is his account of this undertaking, and includes many photographs made by Edward Bierstadt, older brother of the painter Albert Bierstadt. Gorringe studied other obelisk moves, including Fontana's, but instead of relying on men and horses using capstans to turn the obelisk, he designed his own machinery, shown here. The project, completed on January 22, 1881, took fifteen months, including 112 days to move the obelisk from its landing point at 96th Street and the Hudson River to its present location, in Central Park behind the Metropolitan Museum of Art.

the skin and see into the human body. Röntgen's discovery in 1895 of the X ray was probably one of the few scientific advances that had immediate and practical utility; it quickly became indispensable in clinical medicine for diagnosis and treatment.

The revisions of the theories and laws governing electricity, magnetism, and light in the latter half of the nineteenth century were crucial to the work of Einstein and others. The problem concerning the nature of light, however – was it a wave? was it a particle? – remained the subject of a great deal of scientific speculation and experimentation. Quantum mechanics grew out of this work and became in turn the basis for what is popularly known as nuclear or atomic physics. It may seem that the Greeks' definition of physics as the study of the "nature of things" is too simplistic for matters that seem terribly complex and sophisticated to the layman, but it is still appropriate. Even Einstein would have conceded that it was the "nature of things" he was trying to explain when he promulgated his theory of relativity.

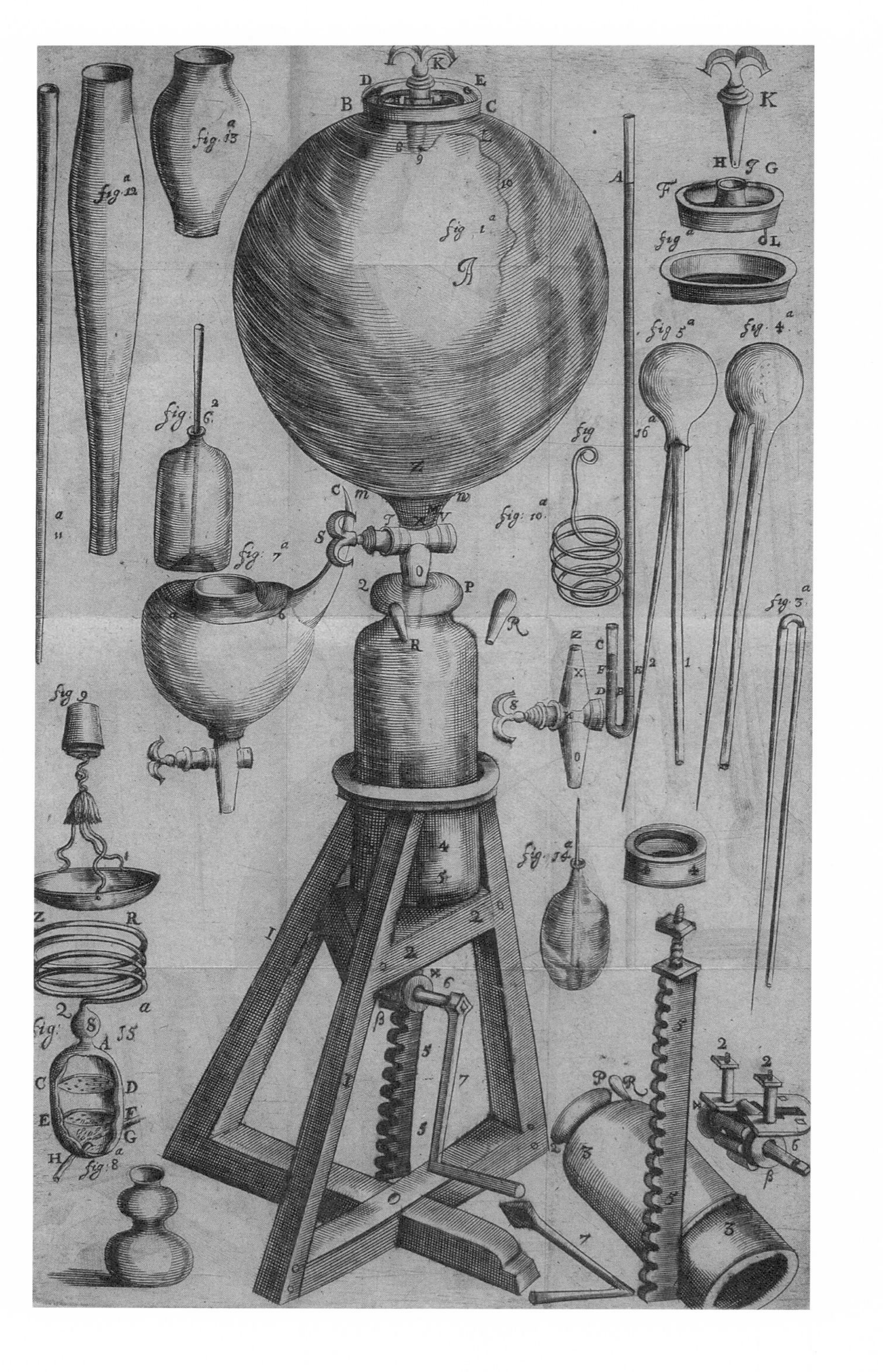

fig. 13 a
fig. 12 a
fig 1 a
fig: 6. 2
fig. 7 a
fig: 10. a
fig 5 a
fig. 4 a
fig. 3 a
16 a
fig 9
fig. 14 a
fig. 15 a
fig. 8 a

Chemistry

CHEMISTRY AROSE FROM a constellation of diverse pursuits. Brewers, distillers, tanners, jewelers, dyers, glassmakers, and miners all used the reaction of matter as a basis for their trades. Alchemists sought material perfection through the preparation of the philosophers' stone, to change base metals to gold, and the Elixir of Life, to create immortality and redemption for humans. Physicians sought a variety of preparations for their patients, including gold. More important than alchemists, "iatrochemists" (from the Greek *iatros,* physician) stressed the importance of chemical remedies, leading to the development of pharmaceutical laboratories in the eighteenth century.

Before the mid-seventeenth century, chemistry, like physics, included the study of all natural phenomena. At that time, Robert Boyle began to lay the foundations of modern chemistry through his work on the structure of matter. Using a modified version of

---

After hearing of Otto van Guericke's invention of the air pump to create a vacuum in 1654, Robert Boyle had improved versions made for himself and began to conduct a series of careful experiments on the physical nature of air. Working with Robert Hooke, who was then in his twenties, Boyle was able to demonstrate many of the properties of air that we now take for granted, including the fact that air is necessary for combustion and respiration. The first edition of his *New Experiments* appeared in 1660; it was followed by a second edition in 1662, which was reprinted in Latin in 1669 and included this plate showing some of his apparatus. It was in the second edition that he presented what he called his "hypothesis," now called Boyle's Law: that the volume of air in a confined space varies inversely with pressure.

"Marie Curie is, of all celebrated beings, the one whom fame has not corrupted."

ALBERT EINSTEIN (1879–1955)

Otto von Guericke's air pump, newly invented in 1654, to create a vacuum, Boyle carried out a series of experiments on the physical nature of air, which led in 1662 to his discovery of the law named after him: the volume of a gas is inversely proportional to the pressure. His experiments on combustion, first carried out with Robert Hooke, his assistant at Oxford, were essential to later attempts to isolate distinct gases. Karl Wilhelm Scheele and Joseph Priestley produced oxygen independently by 1774. Antoine Lavoisier was the first to recognize the implications of the discovery of oxygen. He designated oxygen an element and established the concept of elements as substances that cannot be further broken down. Many of the published works of these men are illustrated by depictions of their apparatus, since they wanted readers to be able to reproduce their results exactly.

In 1808, John Dalton presented the idea that a chemical element was something quantifiable, and his *A New System of Chemical Philosophy* is illustrated with symbols representing each element. But Dalton's symbols were found to be too cumbersome, and five years later, Jöns Jacob Berzelius proposed a system (still in use today) in which the first letter of the Latin name of the element, such as "H" for hydrogen, represents that element.

The modern theory of atomic structure owed a great deal to early work on isolating the elements, but was primarily based on the work done on radioactivity and the discovery of the electron. J. J. Thomson, in 1897, demonstrated that there were negatively charged particles much smaller in mass than an atom of hydrogen. He speculated that these particles were part of all atoms. Pierre and Marie Curie and Henri Becquerel integrated this information into their study of radioactivity.

Throughout the nineteenth and twentieth centuries, chemistry has retained a central position in relation to the other sciences and industry. Chemists today often engage in interdisciplinary and applied research, in new combinations such as biochemistry, photochemistry, and electrochemistry. Although some might see chemistry as once again a part of physics, many chemists consider their discipline to be as autonomous as architecture, using molecules to create structures that obey certain rules.

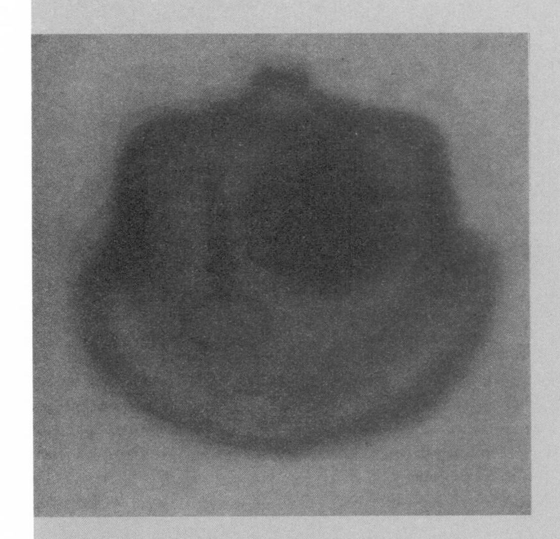

Marie Sklodowska Curie (1867–1934) was one of the most remarkable women of the late nineteenth and early twentieth centuries, and one of the few well-known women scientists of any period. Her life consisted of an amazing litany of "firsts." She was the first woman to obtain a degree in physics and mathematics from the Sorbonne, and to teach there (she was appointed in May 1906 to the chair of physics created in 1904 for her husband, Pierre, and vacated at his death in April 1906). In 1903, only six months after defending her thesis, she became the first woman to be awarded a Nobel Prize, in Physics; she shared that award, for the discovery of radioactivity, with her husband and with Henri Becquerel. In 1911, she became the first Nobel laureate to receive a second Nobel Prize, this time in chemistry for the discovery of radium and polonium. (Henri Becquerel had discovered the radioactivity of uranium in 1896.) And in 1922, she was finally admitted to the Académie des Sciences, established in the seventeenth century, as its first woman member.

ABOVE: *Recherches sur les substances radioactives* (1904) was the second, revised edition of Marie Curie's 1903 doctoral thesis, and presents her research in radioactivity from 1897 to 1904. This image, from the 1904 German edition, demonstrates the effectiveness of passing beta and gamma rays through nonmetallic substances to achieve an effect very much like Röntgen's X ray. Various metallic objects, such as this key and coin in a coin purse, become visible when beta and gamma rays are stopped by the metal.

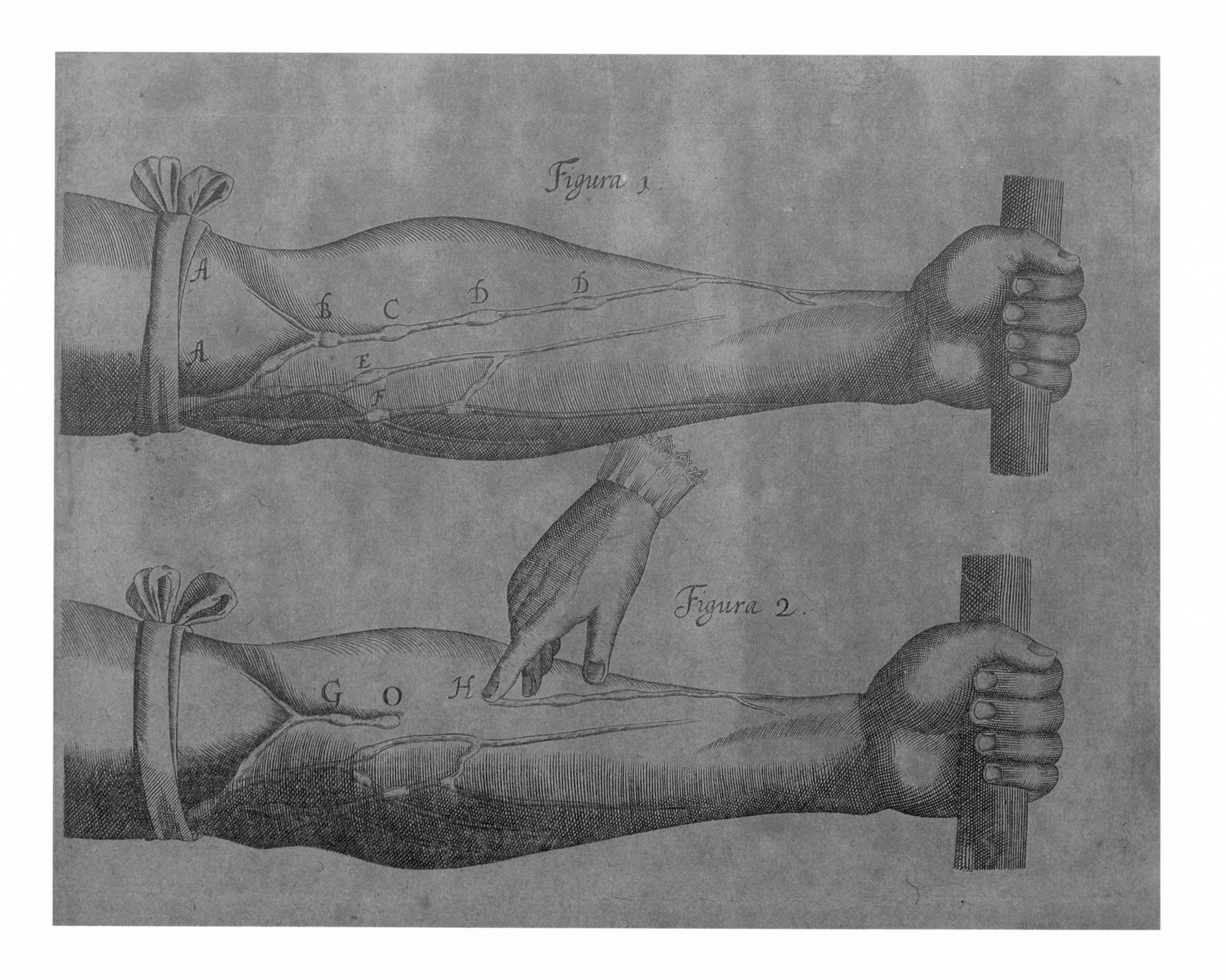
Figura 1.
A
A
B
C
D
D
E
F
Figura 2.
G
O
H

Medicine

EUROPEAN MEDICINE, WHICH began with Hippocrates, maintained the authority of its classical antecedents far longer than any other scientific discipline. By the end of the seventeenth century, Aristotle and Ptolemy had been definitively overturned in the physical sciences, but medical practitioners still perceived disease as an imbalance in the bodily humours and routinely bled their patients if they were too sanguineous or dosed them if they were too bilious.

However, the early seventeenth century saw the publication of the most important book in the history of medicine: William Harvey's *Exercitatio anatomica de motu cordis et sanguinis in animalibus* (1628), describing Harvey's discovery of the circulation of the blood. Although its immediate effect on the practice of clinical medicine was negligible, its influence on scientific thought was profound; it was as revolutionary for the life sciences as Newton's *Principia* (1687) would be for physical science toward the end of the century.

That the blood moves through the body is an observation that may have occurred to almost anyone who has ever seen a severed

In his *De motu cordis* (1628), William Harvey demonstrated the continuous circulation of the blood. The two plates illustrating the valves of the veins and the veins (one of which is shown here) were copied from illustrations in the anatomical work of Caspar Bauhin. Bauhin's illustrations were in turn based on the illustrations for the work of Girolamo Fabrici, the discoverer of the valves of the veins.

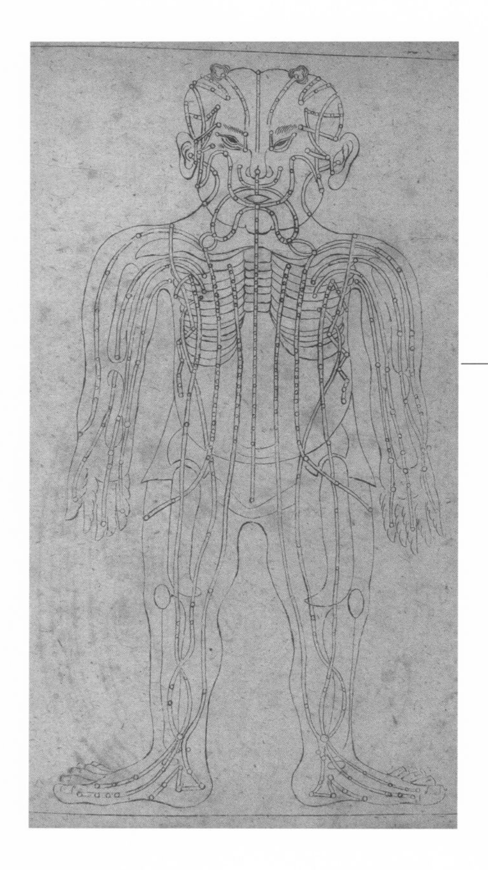

Andreas Cleyer's *Specimen medicinae sinicae* (1682) was the first illustrated work on Chinese medicine to be published in the West. Cleyer, a physician with the Dutch East India Company who worked in Java and later in Japan, is given credit for the text, but he actually translated the work of the Polish Jesuit Michael Boym, who relied on various Chinese authorities. The engravings were probably copied from Chang Chieh-pin's *Lei ching* (1624). They include several acupuncture charts, the first to be published in the West, such as this one; mistaken for anatomical charts by Cleyer's contemporaries, they generated much misguided criticism of the work.

artery. Even the notion that blood moves in a definite path may have been entertained as a possibility by the ancient Egyptians and Greeks, whose medicine was rather advanced. Girolamo Fabrici, a professor of medicine at the University of Padua with whom William Harvey had studied, discovered the valves of the veins, but it was left to Harvey to propose a single circulatory system with the heart, a muscular pump, at its center. Like Newton, Harvey succeeded in establishing experiment and observation as the basis for his scientific advances and supported his subsequent assertions with precise mathematical measurements. Through his efforts, quantitative analysis became a means by which life processes could be determined and described.

But even though microscopy, experimentation, dissection, and various work in the physical sciences had explained a great deal about human anatomy and physiology, the application of that knowledge to the practice of clinical medicine would come only later. For example, germ theory and antisepsis were late developments: in 1843, Dr. Oliver Wendell Holmes occasioned violent protest by suggesting at a meeting of the Boston Society for Medical Improvement that physicians wash their hands and change their clothes before delivering

The forty magnificent full-page copperplate engravings of human figures in Bernardino Genga's *Anatomia* (1691) were intended primarily for the use of artists and sculptors. Probably engraved by François Andriot from the drawings of Charles Errard, the illustrations depict the human body in a variety of poses, some with dissection and some undissected. Some of the figures are anatomical representations of works of classical sculpture, such as the Farnese Hercules, the Borghese Faun, and, shown here, the Laocoön without his sons or the snake. The work is one of the most beautiful of the great books of anatomy and stands as one of the best examples of the great age of copperplate engraving in the seventeenth century.

Desiré Magloire Bourneville and Paul Régnard's case studies of hysteria and epilepsy were published as *Iconographie photographique de la Salpêtrière* (1877–80). Although the second and third volumes were illustrated with collotype plates, the first issue of Volume I was illustrated with forty albumen print photographs by Régnard mounted on cards; this one shows a woman in the last stage of a grand mal, when the seizure has subsided. (Later issues of Volume I replaced the photographs with collotype plates.) Bourneville was Jean Martin Charcot's assistant at La Salpêtrière in Paris from 1870 to 1879. In 1882, Charcot, the founder of the clinical specialty of neurology, opened his neurological clinic at Salpêtrière, where his students included Sigmund Freud.

babies. During the American Civil War, the mortality rate from gangrene and other infections among casualties far exceeded battle deaths. It was not until World War I that antisepsis finally became routine enough for surgeons to scrub and don special surgical garb before performing operations. And only with the discovery of penicillin, and consequent antibiotics, in the twentieth century were physicians finally able to actually cure a bacterial infection.

Although the practice of clinical medicine remained static for some time, medical illustration did not. By the end of the seventeenth century, observation and experimentation, combined with advances in science, microscopy, perspective, and artistic technique, had resulted in markedly improved anatomic illustrations. To the artists whom anatomists employed as illustrators, beauty was as great a concern as was accuracy of depiction. Works such as Bernardino Genga's *Anatomia* (1691) were intended primarily for the use of artists and sculptors, rather than for medical practitioners.

PLATE 1. Joannes de Sacro Bosco, *Comptus, quadrans, de sphaera, algorismus* (France, ca. 1275)

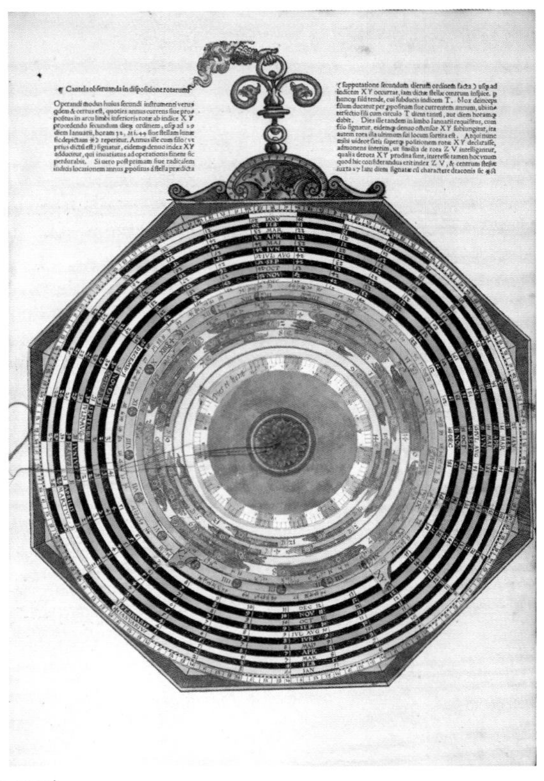

PLATES II–III. Petrus Apianus, *Astronomicum caesareum* (Ingolstadt, 1540)

PLATE IV. Andreas Vesalius, *De humani corporis fabrica libri septem* (Basel, 1543)

PLATE V. Étienne Léopold Trouvelot, *Astronomical Drawings* (New York, 1882)

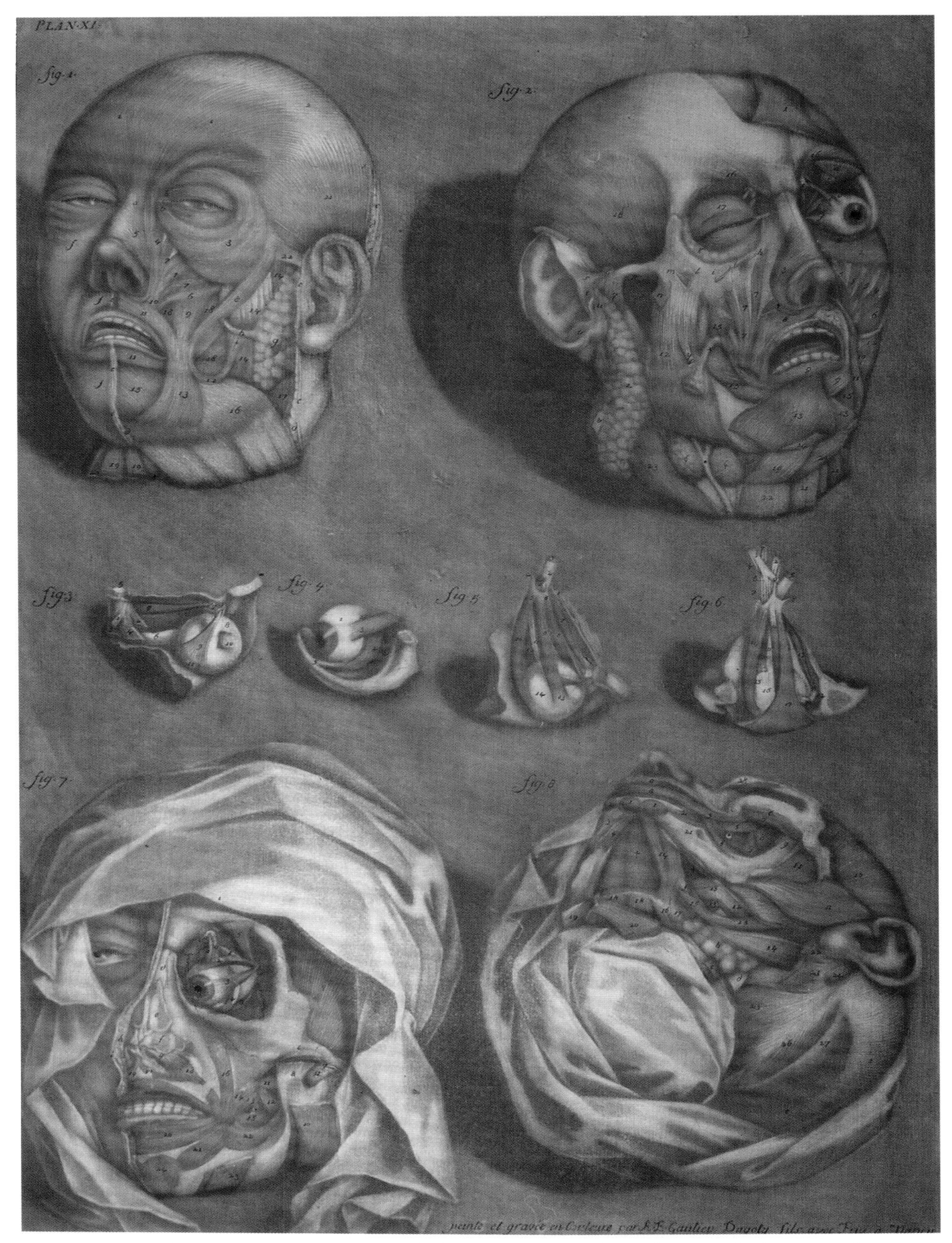

PLATE VI. Arnaud Éloi Gautier-d'Agoty, *Cours complet d'anatomie peint et gravé en couleurs naturelles* (Nancy, 1773)

PLATE VII. Maria Sibylla Merian, *Metamorphosibus insectorum Surinamensium . . . Dissertation sur la generation et les transformations des insectes de Surinam* (The Hague, 1726)

PLATE VIII. Anna Atkins, *Photographs of British Algae: Cyanotype Impressions* (Halstead Place, Sevenoaks, England, 1843–53)

PLATE IX. Edward Lear, *Illustrations of the Family of Psittacidae, or Parrots* (London, 1832)

"At least we have made such progress that we do see we can to a great extent so record that our successors yet unborn can also see; and it is owing to this fact that a part of the medical literature of the last quarter of the nineteenth century will be more valuable than all that has preceded it."

John Shaw Billings (1838–1913)

Photography brought its own concerns and applications. Where an artistic rendering would allow for an image incorporating knowledge gleaned from numerous anatomic dissections, a photograph could show only a specific image at a particular moment in time. The pedagogic utility of the former was far superior, and to this day anatomical figures are usually rendered artistically rather than photographically. On the other hand, photographs were employed in neurological texts such as the *Iconographie photographique de la Salpêtrière* (1877–80) to chronicle the progression of a grand mal seizure in a patient. They were also used to demonstrate particular characteristics of neurological impairment, or to trace the path of low-level electric current applied by electrodes strategically placed on the head, face, or torso.

The effects of treatment and new clinical or surgical methodologies were also best illustrated through photographs; there exist numerous clinical photographs of amputees, gunshot wounds, and various medical apparatus from the American Civil War. Photographs were useful to both the practitioner and the theorist, and the demand for them increased as the technology improved. The problem for printers and publishers then became finding a way to reproduce photographs in books cheaply and easily enough to make such publications profitable; the first photographs of various bacteria in color did not appear in medical texts until well into the twentieth century.

Natural History

FROM MEDIEVAL TIMES, naturalists have sought to describe all living things in order to find their particular uses for mankind. During the Renaissance, the natural world began to be studied apart from its utilitarian value. The process of direct observation established by Vesalius in 1543 for human anatomy, and by Brunfels and Fuchs for botany in 1530 and 1542, was aided by many other factors. The precursors of today's museums, cabinets of natural history such as those of Ferrante Imperato in the sixteenth century, and Ole Worm in the seventeenth century, collected specimens of the known and unknown world. These encyclopedic museums served not only to provide pleasure to the visitor but also began the process of the systematic classification of the natural world. Imperato's system, for example, was based partly on usage and partly on the behavior of materials: for the various kinds of earth, he used the divisions *agricolarum* (for farming), *plasticorum*

With *Études sur les glaciers* (1840), Louis Agassiz became the founder of glacial geology. While observing various glaciers near Chamonix and in the Rhone Valley, including this one in Zermatt, Agassiz realized that the smooth rock faces must have been created by ice flow, not water flow. Further research led him to the discovery that most of Europe, North Asia, and North America had at one time been covered by ice, during a period he named the Ice Age. Agassiz rejected evolutionary biology, but he was the leading proponent of the careful study of the natural world, combining fieldwork with research, teaching, lecturing, and publishing, and many of his discoveries provided important information for the work of Lyell and Darwin.

The double-page frontispiece to pharmacist Ferrante Imperato's *Dell'historia naturale* (1599) shows his library in Naples. A cabinet of natural wonders, Imperato's library, which operated from 1599 until 1670, long after the collector's death, included not only books, but also scientific specimens, instruments, furniture, and other objects of interest to their collector.

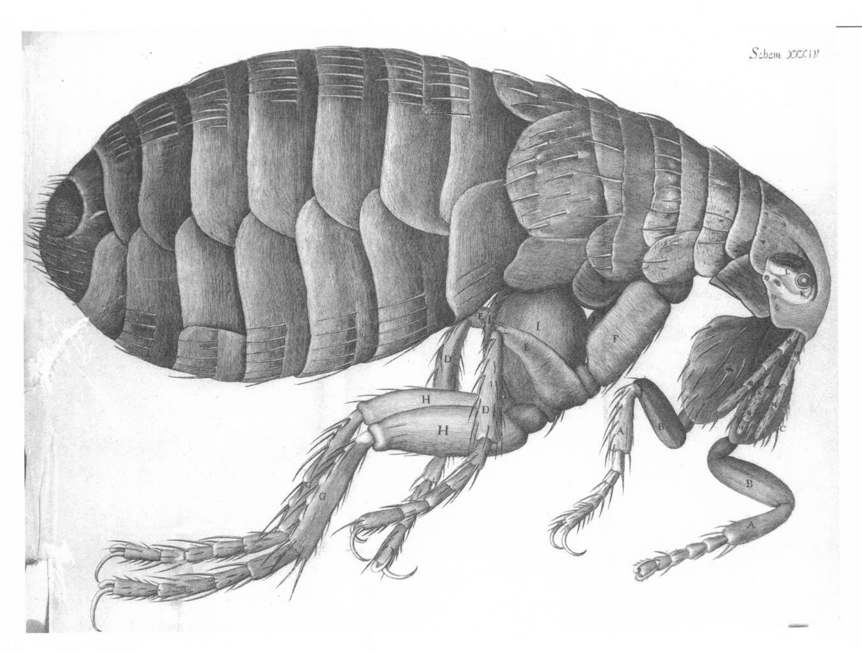

Although Robert Hooke's *Micrographia* (1665) was not the first book to reproduce microscopical observations, it was the first devoted to them, and the first to reproduce observations made with Hooke's newly perfected compound microscope. It was also the first book to include detailed illustrations for each section of text, drawn and in some cases also engraved by the author himself. The impact of this book was tremendous – indeed, his nearly 16-inch-long engravings of the flea, shown here, and the louse are reported to have changed the hygiene habits of many of his contemporaries – and Hooke's name became synonymous in the public mind with microscopical observations.

(for building), *fusorum* (for metallurgy), *pictorum* and *fullonum* (for artists and dyers), and *medicorum* (for health).

Explorers brought to Europe thousands of previously unknown species of plants and animals, further undermining the classical Greek and Roman authorities, such as Aristotle and Dioscorides. Maria Sibylla Merian, one of the world's great entomologist-artists, became interested in the insects of the Dutch colony of Surinam after seeing specimens of Surinam insects in Dutch collections. In 1699, after leaving her husband and selling a collection of her paintings to finance the voyage, she and her youngest daughter, Dorothea Maria, went to Surinam, spending two years drawing and painting insects, flowers, fruit, and reptiles. After returning to Amsterdam, she produced her greatest work, *On the Metamorphosis of the Insects of Surinam* (1705). New classification schemes were developed so that naturalists could begin to make sense of the rapidly expanding natural world. These

"Darwin's book has made a greater sensation than any strictly scientific book that I remember. It is wonderful how much it has been talked about by unscientific people; talked about of course by many who have not read it, and by some I suspect, who have read without understanding it, for it is a very hard book."

CHARLES BUNBURY (1809–1886)

included alphabetical schemes such as that of Konrad Gesner and more complicated systems based on individual characteristics of specimens, such as those of Brunfels, Guillaume Rondelet, Pierre Belon, and Thomas Willis. Later, Linnaeus's work led to Darwin's breakthrough system based on genealogical descent.

Exploration also brought a new awareness of the geographic distribution of plants and animals. If all creatures had come out of the Ark after the biblical flood, how could they have found their way to such places as America and Australia? And what could account for the great differences as well as similarities of species of living things found throughout the globe? These questions led to the work of Darwin and others in developing the theory of evolution.

In addition, Darwin's recognition of the oneness of nature, as represented by his tree of life in *On the Origin of Species* (1859), stands as a landmark concept, creating a model for the understanding of the historical relationships of all life forms. Equal to Copernicus's removal of the Earth as the center of our universe, Darwin's removal of humankind as the apex of creation allowed natural history to develop as a truly modern science.

Robert Hooke's invention of the compound microscope, the first to resemble today's microscopes, allowed him to produce some of the finest microscopic observations ever made. His examination in the *Micrographia* (1665) of the structures of many common things, such as woven cloth, fish scales, feathers, sponges, the sting of a bee, the eye of a fly, and human hair, opened whole new worlds of scientific discovery. More accurate depictions of the natural world were also made possible by developments in illustration techniques, including nature printing, made with actual specimens, and photography, beginning with the cyanotypes of Anna Atkins, the first woman photographer.

From its medieval beginnings as the study of animals, plants, and minerals, natural history has grown to encompass a number of fields in the life and earth sciences, including botany, zoology, entomology, ornithology, biogeography, ecology, geology, and paleontology.

Charles Darwin (1809–1882) was not the first person to theorize about evolution, but he was the first to combine the accumulated discoveries and observations in natural history with his own into a comprehensive theory of nature. His grandfather, Erasmus Darwin (1731–1802), had proposed the possibility of the transmutation of living forms. His friend and confidant, Sir Charles Lyell (1797–1875), although not an evolutionist, also contributed to evolution theory by establishing that the Earth was very old and that organic change could therefore occur slowly over a vast period of time. Lyell also dismissed the idea of the catastrophic extinction of animal forms on a worldwide scale and asserted that the action of natural forces, over enormous periods of time, was sufficient to account for the phenomena found in rocks, such as fossils.

Before his famous voyage aboard H.M.S. *Beagle* in 1831–36, to survey portions of the coast of South America and to visit some of the Pacific islands, Darwin was a country gentleman who had failed at medicine but who, it was thought, might be successful as a country parson. His mind, which had not been subjected to the rigors and prejudices of a nineteenth-century classical education, was free to observe and notate the bits of flora and fauna that he found halfway around the world and then develop a theory whose implications were revolutionary.

PICTORES OPERIS,

Heinricus Füllmaurer. Albertus Meyer.

SCVLPTOR

Vitus Rodolph. Speckle.

Illustration Processes

All book illustration processes can be divided into four basic groups: relief, intaglio, and planographic printing, and photography. In the 1960s, when text as well as illustrations began to be reproduced photographically, offset lithography, a photolithographic printing process, took over commercial book production and held the field until today's digital revolution.

Relief Printing

Relief printing is the oldest of the printed illustration processes. The woodcut, in which a knife is used on the plank side of a piece of wood to carve away everything except the lines to be printed, was invented in China during the eighth century, and reached Europe around 1400. With the invention of the printing press and movable type around 1455, the woodcut became the primary means of printing illustrations in books, since the lines of the image stood up in relief in the same way as did the type, and they could be printed together on the same press.

In *De historia stirpium commentarii* (1542), Leonhart Fuchs presented a compilation of the known plants that were useful as drugs and herbs, illustrated with woodcuts intended to be both accurate and beautiful. To emphasize the importance of the process of creating the illustrations, Fuchs included portraits of the three artists as a kind of colophon to the book. Albrecht Meyer (upper right) is shown drawing the design of a pink from the plant itself. Heinrich Füllmaurer (upper left) is shown transferring the drawing to the woodblock. Veit Rudolf Speckle, the woodcut artist or "sculptor," is shown below.

CALIPER
Various Processes

GRAVER
Engraving

WOOD-ENGRAVED BLOCK

Used by such artists as Albrecht Dürer and those of the school of Titian who worked for Vesalius, the woodcut lost its primitive appearance. But as woodcuts became more sophisticated, the process of cutting away the nonprinting areas became more tedious. By the seventeenth century, intaglio had generally replaced the woodcut.

Toward the end of the eighteenth century, the relief process experienced a revival that continues to this day in small and fine press printing. The English engraver Thomas Bewick perfected a method of using engraving tools on the end grain of a hard wood, such as boxwood, to cut away the nonprinting surfaces, while creating very fine details. This technique led to an explosion of illustrated books, magazines, and newspapers beginning in the mid-nineteenth century. Alexander Anderson, who followed in the footsteps of Bewick, is considered the father of American wood engraving. After the invention of photography, some wood engravers were able to mimic the appearance of a photograph. With the invention of the relief halftone process in the 1880s, printing blocks could be made directly from photographs, and wood engraving became obsolete for commercial publishing.

Intaglio Printing

Intaglio (Italian, "to incise") includes engraving, etching, and mezzotint, among other techniques. Reversing the relief process, in intaglio the artist cuts the lines to be printed, rather than cutting away the

DAUBER
Various Processes

NEEDLE
Engraving

ROCKER
Mezzotint

nonprinting surfaces. Although it is an ancient process found in cave paintings and related to the engraving of seals, intaglio did not come into use in Europe for printing illustrations until the fifteenth century. Engraving allowed the scientific or medical artist to create a more precise and detailed line in a metal plate – copper at first, but later steel – than was possible in relief. Although not new when used by Robert Hooke for his *Micrographia* (1665), engraving perfectly suited his exacting work with his newly perfected compound microscope. However, intaglio printing requires much more pressure than relief, since the ink is held in recessed grooves instead of on the surface of the plate, and so illustrations could not be printed on the same press as the text. When printed on separate sheets (which came to be called "plates"), the illustrations could be bound in to face corresponding text, as in *Micrographia*, or placed together at the end for easy reference.

Cutting into metal requires great strength and skill, and so a less arduous process, etching, was developed in the early sixteenth century. Etching uses acid to cut into the plate, which allows the artist much greater freedom to create a line. Easily combined with engraving, it became the intaglio process most favored by artists. Since pure engraving and pure etching became uncommon, the term "line engraving" is generally used to describe the commercial prints of the eighteenth and nineteenth centuries that combined etching and engraving.

Another type of intaglio process, the mezzotint (Italian, "half-tint") allowed an artist to create the middle range of tones between black and white. In a laborious process, the overall ground is laid down with a rocker, and the end result has a velvety quality. Although used by Gautier-d'Agoty for his anatomical atlas (1773), mezzotint was found to be too imprecise for scientific and medical illustration.

Planographic Printing

Planographic printing is printing from a flat surface, or plane. Now considered the first and most important method of planographic printing, lithography was invented in Germany in 1798 by Aloys Senefelder. Lithography is based on the chemical principles that oil and water do not mix but will attract like substances, and that both will adhere to a porous ground, such as a stone. The process came into widespread use in the 1820s as commercial printers and artists realized that images could be drawn as easily on stone as on paper, and that the stones could be reused. All methods of drawing could be used on the stone, including pen-and-ink, chalk, or crayon, and by about 1830, Charles Hullmandel had invented a watercolor-like wash, applied with a brush, that provided tints known as lithotints. Only after the development of cheaper printing methods such as wood engraving and lithography did it become possible to print inexpensive illustrated textbooks, such as Gray's *Anatomy* (1858) for medicine. Although lithographs could be produced more easily, and generally more cheaply, than relief or intaglio illustrations, color was still expensive since it had to be applied by hand. For example, to create his *Parrots* (1832), Edward Lear drew the black portion of the images on stone. The color was then added by hand to Lear's specifications.

During the nineteenth century, lithographers perfected the art of printing in color by using multiple stones to achieve very complex colored images through a process known as chromolithography. Cheap color printing was then available for the first time in the history of printing. Julius Bien's unfinished American edition of Audubon's elephant folio *Birds of America* (1860–61) was the most ambitious chromolithographic book project undertaken in America.

Photography

Credit for the invention of photography is attributed variously to Joseph Nicéphore Niépce, Louis Daguerre, and William Henry Fox Talbot. Niépce made the first permanent photographic images in 1826–27, using bitumen-coated pewter. He then worked with Daguerre on a process that became known as the daguerreotype, announced early in 1839. No negative was involved; the image was created through the action of light on silver-coated copper treated with mercury vapor to make it light sensitive. Shortly thereafter, Talbot announced his discovery of a rival process using a negative, which allowed the image to be printed repeatedly, unreversed, on another medium such as paper. This process became the standard for photography until the recent advent of digital photography.

Most early use of photography in books required that separately printed photographs be inserted into each volume of an edition by hand, a costly process. With the invention of wood engraving, skilled artists could create the appearance of a photograph. By the 1880s, a much cheaper method had come into use: the halftone photomechanical process used a screen to transform the photographic image into dots. Depending on how the dots were reproduced, halftones could be printed by relief or planographic printing methods. When combined with the use of the three primary colors, halftones drove out chromolithography for most color printing by the beginning of the twentieth century.

The first photographs of the moon were taken using the daguerreotype process, and photographs were also taken of the Sun's corona during the 1840s. Anna Atkins, the first woman photographer, photographed specimens of plants for her *British Algae* of 1843–53. But a photograph is often the least informative means of illustration since it presents too much information about a specific subject. In anatomy and in botany, the old methods of drawing are still used to illustrate general principles, since photography can show only a specific human body or plant specimen.

Checklist of the Exhibition

In this checklist, capitalization of titles has been regularized; place names have been anglicized; and publishers' names have been standardized. Items illustrated in this publication are indicated by an asterisk (*); the page or plate number is given in brackets at the end of each entry. The following abbreviations are used to indicate the location of each item.

Collections of The New York Public Library

ARENTS: BOOKS IN PARTS	Arents Collection of Books in Parts, Humanities and Social Sciences Library
ARENTS: TOBACCO	Arents Tobacco Collection, Humanities and Social Sciences Library
BERG	Henry W. and Albert A. Berg Collection of English and American Literature, Humanities and Social Sciences Library
BURDEN	Carter Burden Science Fiction Collection, Henry W. and Albert A. Berg Collection of English and American Literature, Humanities and Social Sciences Library
DREXEL	Drexel Collection, Music Division, The New York Public Library for the Performing Arts
DUYCKINCK	Duyckinck Collection, Print Collection, Miriam and Ira D. Wallach Division of Art, Prints and Photographs, Humanities and Social Sciences Library
GRD	General Research Division, Humanities and Social Sciences Library
JWD	Dorot Jewish Division, Humanities and Social Sciences Library
MAP	Map Division, Humanities and Social Sciences Library

MSS	Manuscripts and Archives Division, Humanities and Social Sciences Library
ORT	Oriental Division, Humanities and Social Sciences Library
PARSONS	Parsons Collection, Science, Industry and Business Library
PFORZHEIMER	The Carl H. Pforzheimer Collection of Shelley and His Circle, Humanities and Social Sciences Library
PHOTOGRAPHY	Photography Collection, Miriam and Ira D. Wallach Division of Art, Prints and Photographs, Humanities and Social Sciences Library
PRINTS	Print Collection, Miriam and Ira D. Wallach Division of Art, Prints and Photographs, Humanities and Social Sciences Library
RBK	Rare Books Division, Humanities and Social Sciences Library
ROSIN	Axel Rosin Collection of German Literature, Rare Books Division, Humanities and Social Sciences Library
SC	Schomburg Center for Research in Black Culture
SCHLOSSER	Leonard B. Schlosser Collection on the History of Papermaking, Print Collection, Miriam and Ira D. Wallach Division of Art, Prints and Photographs, Humanities and Social Sciences Library
SIBL	Science, Industry and Business Library
SLAUGHTER	Lawrence H. Slaughter Collection, Map Division, Humanities and Social Sciences Library
SPENCER	Spencer Collection, Humanities and Social Sciences Library
STUART	Robert L. Stuart Collection, Rare Books Division, Humanities and Social Sciences Library
WHEELER	Wheeler Collection, Rare Books Division, Humanities and Social Sciences Library

Other Collections

MERTZ	The LuEsther T. Mertz Library, The New York Botanical Garden
NYAM	The New York Academy of Medicine

The Medieval Worldview

Anglicus, Bartholomaeus (fl. 13th century). *De proprietatibus rerum.* [Westminster]: Wynkyn de Worde, [1495]. RBK

Antichrist. *Von dem Endkrist.* Strasbourg?, ca. 1482. RBK

*Apianus, Petrus (1495–1552). *Astronomicum caesareum.* Ingolstadt: for the author, 1540. RBK; Purchased from the Bequest of Alexander Maitland [PLATES II–III]

Berengario da Carpi, Jacopo (ca. 1460–ca. 1530). *Isagoge breves perlucide ac uberime in anatomiam humani corporis.* Venice: Bernardino di Vitale, 1535. NYAM

Biringuccio, Vannoccio (1480–1539?). *De la pirotechnia.* Venice: Venturino Rossinello, 1540. PARSONS

Brissot, Pierre (1478–1522). *Apologetica disceptatio, qua docetur per quae loca sanguis mitti debeat in viscerum inflammationibus, praesertim in pleuritide.* Paris: Simon de Colines, 1525. NYAM

Brunfels, Otto (1488–1534). *Herbarum vivae eicones.* Strasbourg: Johann Schott, 1530. SPENCER

Brunschwig, Hieronymus (ca. 1450–ca. 1512). *Buch der chirurgia.* Augsburg: Johann Schönsperger, 1497. NYAM

———. *Chirurgia, das ist, Handwürckung der Wundartzney.* Augsburg: Alexander Weyssenhorn, 1534. SPENCER

———. *Kleines Distillierbuch.* Strasbourg: Johann Grieninger, 1500. NYAM

Chauliac, Guy de (ca. 1300–1368). *Chirurgia.* England?, late 14th century? NYAM

*Copernicus, Nicolaus (1473–1543). *De revolutionibus orbium coelestium.* Nuremberg: Johannes Petreius, 1543. RBK; from the Astor Library [PAGE 18]

Dati, Gregorio (1362–1436). *La sfera.* North Italy, ca. 1450–65. MSS; Purchased from the Bequest of Alexander Maitland

Dryander, Johann (1500–1560). *Anatomiae, hoc est, corporis humani dissectionis pars prior.* Marburg: Eucharius Cervicorn, 1537. NYAM

*Dürer, Albrecht (1471–1528). *Unterweysung der Messung.* Nuremberg: for the author, 1525. SPENCER [PAGE 8]

Estienne, Charles (1504–ca. 1564). *La dissection des parties du corps humain.* Paris: Simon de Colines, 1546. RBK; John Shaw Billings Memorial Collection

*Euclid (fl. ca. 300 B.C.E.). *Elementa geometricae.* Venice: Erhard Ratdolt, 1482. RBK; John Shaw Billings Memorial Collection [PAGE 28]

*Fuchs, Leonhart (1501–1566). *De historia stirpium commentarii insignes.* Basel: Michael Isingrin, 1542. SPENCER [PAGE 56]

Gaffurius, Franchinus (1451–1522). *Theorica musicae.* Milan: Philip de Mantegat, Cassan, for Johannes Petrus de Lomatio, 1492. DREXEL; from the Lenox Library

Gersdorff, Hans von (ca. 1455–1529). *Feldtbüch der Wundartzney.* Strasbourg: Johannes Schott, 1517. NYAM

Hortus sanitatis. *Ortus sanitatis de herbis et plantis.* Strasbourg: Johann Prüss, 1496. SPENCER

Hutten, Ulrich von (1488–1523). *Von der wunderbarliche Artzney des holtz Guaiacum genant.* Strasbourg: Johann Grieninger, 1519. RBK; Purchased from the Bequest of Alexander Maitland

Hyginus, Caius Julius (28 B.C.E.–10 C.E.). *De sideribus tractatus.* Padua, ca. 1465–70. SPENCER

———. *Poeticon astronomicon.* Venice: Erhard Ratdolt, 1482. RBK; Purchased from the Trust Fund of Lathrop Colgate Harper

Jābir ibn Ḥaiyān, al-Tartūsī (fl. 776). . . . *De alchemia.* Nuremberg: Johannes Petreius, 1541. RBK

★Ketham, Joannes de (15th century). *Fasciculo di medicina.* Venice: Joannes and Gregorius de Gregoriis, 1493. SPENCER [PAGE 16]

———. *Fasciculus medicinae.* Venice: Joannes and Gregorius de Gregoriis, 1495. NYAM

Manṣūr ibn Muḥammad ibn Aḥmad ibn Yūsuf ibn Faqīh Ilyās (fl. 15th century). *Al-Tashrīḥ bi-al-Taṣvīr [Anatomy with Pictures].* Persia, ca. 1675. SPENCER

Megenberg, Konrad von (1309–1374). *Buch der Natur.* Augsburg: Johann Bämler, 1478. NYAM

Pacioli, Luca (d. ca. 1514). *Summa de arithmetica, geometria, proportioni: et proportionalià. Novamente impressa.* Venice: Paganinus, 1494. RBK

Paracelsus (1493–1541). *Der grossen Wundartzney: das erst Buch.* Augsburg: Heynrich Steyner, 1536. NYAM

Ptolemaeus, Claudius (fl. 2nd century C.E.). *Almagestum.* Venice: Peter Liechtenstein, 1515. RBK; from the Astor Library

Regiomontanus [Johannes Müller] (1436–1476). *Epytoma . . . in almagestum Ptolemaei.* Venice: Johannes Hamman, 1496. RBK

Reisch, Gregorius (d. 1525). *Margarita philosophica.* Freiburg: Johann Schott, 1503. RBK

———. *Margarita philosophica.* Strasbourg: Johann Schott, 1504. RBK; from the Astor Library

Rösslin, Eucharius (d. 1526). *Der Schwangerenn Frauwen und hebammen Rosengarte.* Augsburg: Heinrich Steyner, 1530. NYAM

Ryff, Walther Hermann (d. 1548). *Anatomica omnium humani corporis partium descriptio.* Paris: Christian Wechel, 1543. NYAM

★Sacro Bosco, Joannes de (fl. 1230). *Comptus, quadrans, de sphaera, algorismus.* France, ca. 1275. MSS; Gift of Alexander Maitland [PLATE I]

———. *Tractatum de spera.* Venice: Florentius de Argentina, 1472. RBK; Purchased from the Bequest of Alexander Maitland

Stoeffler, Johann (1452–1531). *Calendarium romanum magnum.* Oppenheim: Jacob Köbel, 1518. RBK; Gift of Alexander Maitland

al-Ṣūfī, 'Abd al-Raḥmān ibn 'Umar (d. 984). *Tarjumah-i Ṣuwar al-kawākib [A Translation of The Patterns of the Stars].* Mashhad?, 1630–33. SPENCER

Tartaglia, Niccolò (1506–1557). *Nova scientia inventa.* Venice: Stephano da Sabio, 1537. PARSONS

Valturio, Roberto (d. 1483). *De re militari.* Verona: Joannes Nicolai de Verona, 1472. SPENCER

★Vesalius, Andreas (1514–1564). *De humani corporis fabrica libri septem.* Basel: Johannes Oporinus, 1543. PRIVATE COLLECTION; Dedication copy presented by Vesalius to Charles V (1500–1558), Holy Roman Emperor, hand-colored throughout and bound in royal purple silk velvet [PLATE IV; photo courtesy of Christie's, New York]

———. *De humani corporis fabrica libri septem.* Basel: Johannes Oporinus, 1555. BERG

———. *Epitome.* Basel: Johannes Oporinus, 1543. NYAM

★———. *Icones anatomicae, tabulae selecta.* Munich: The Bremer Presse for The New York Academy of Medicine and University of Munich Library, 1935. NYAM [PAGE 21]

Vitruvius Pollio, Marcus (fl. first century B.C.E.). *De architectura.* Como: Gotardo de Ponte, 1521. RBK

Astronomy

Bayer, Johann (fl. 1600). *Uranometria.* Ulm: Christopher Mangus, 1603. SIBL; from the Astor Library

★Bode, Johann Elert (1747–1826). *Uranographia.* Berlin: for the author, 1801. SIBL; from the Astor Library [PAGE 25]

Brahe, Tycho (1546–1601). *Astronomiae instauratae mechanica.* Wandsbek: Philip de Ohr, 1598. RBK; from the Astor Library

———. *[Learned Tico Brahae His Astronomicall Coniectur].* London: Michaell Sparke and Samuell Nealand, 1632. RBK

*Case, John. *Sphaera civitates, hoc est reipublicae recte ac pie secundum leges administrandae ratio, cuius ingentem utilitatem.* Frankfurt: Johann Wechel, 1589. RBK [PAGE 22]

Cellarius, Andreas. *Harmonia macrocosmica, sev, Atlas universalis et novus.* Amsterdam: J. Jansson, 1661. MAP

Chiyya, Abraham ben (ca. 1065–ca. 1136). *[Safer Tsurot Ha-aretz] Sphaera mundi.* Basel: Henry Petrus, 1546. JWD; Purchased from the Jacob H. Schiff Fund

Flamsteed, John (1646–1719). *Atlas céleste . . . troisième édition.* Paris: Delamarche, 1795. SIBL; from the Astor Library

———. *Historiae coelestis britannicae.* London: H. Meere, 1725. SIBL; from the Astor Library

Galileo Galilei (1564–1642). *Sidereus nuncius.* London: Jacob Flesher, 1653. SIBL; from the Astor Library

Hevelius, Johannes (1611–1687). *Selenographia; sive, lunae descriptio.* Danzig: Hünefeld, 1647. SIBL

Kepler, Johannes (1571–1630). *Astronomia nova.* [Heidelberg]: G. Vögelinos, 1609. RBK

———. *De stella nova in pede serpentarii.* Prague: Paul Sess, 1606. SIBL; from the Astor Library

———. *Harmonices mundi libri V.* Linz: Johannes Plank for Gottfried Tampach, 1619. DREXEL

———. *Prodromus dissertationum cosmographicarium, continens mysterium cosmographicum.* Tübingen: George Gruppenbach, 1596. RBK; from the Astor Library

———. *Tabulae rudolphinae.* Ulm: Jonas Saur, 1627–30. SIBL

Loewy, Maurice (1833–1907), and Pierre Henri Puiseux (1855–1928). *Observatoire de Paris . . . Atlas photographique de la lune.* Paris: Imprimerie Nationale, Fillon & Heuse, 1896–1904. PHOTOGRAPHY

Mayer, Tobias (1723–1762). *Opera inedita.* Gottingen: J. Christian Dieterich, 1775. SIBL; from the Astor Library

Schiller, Julius (d. 1627). *Coelum stellatum christianum concavum.* Augsburg: Andre Apergeri, 1627. SPENCER

Schmidt, Johann Friedrick Julius (1825–1884). *Charte der Gebirge des Mondes.* Berlin: Dietrich Reimer, 1878. SIBL; Gift of Mrs. Henry Draper

*Trouvelot, Étienne Léopold (1827–1895). *Astronomical Drawings.* New York: Charles Scribner's Sons, [Reproduced from the original drawings by Armstrong & Company, Riverside Press, Cambridge, Massachusetts], 1882. SIBL; from the Astor Library [PLATE V]

Mathematics

Abbott, Edwin Abbott (1838–1926). *Flatland, A Romance of Many Dimensions, With Illustrations by the Author, A Square.* London: Seeley & Co., 1884. BERG

Apollonius of Perga (2nd half third century B.C.E.–early second century B.C.E.). . . . *Conicorum.* Florence: Joseph Cocchini, 1661. SIBL; from the Astor Library

Archimedes (ca. 287–212 B.C.E.). . . . *Archimedis Syracusani philosophi ac geometrae excellentissimi opera . . . omnia.* Basel: Johan Herwagen, 1544. PARSONS

Euclid (fl. ca. 300 B.C.E.). *The Elements of Geometry.* London: John Daye, 1570. STUART

———. *The First Six Books of The Elements of Euclid in which coloured diagrams and symbols are used instead of letters for the greater ease of learners by Oliver Byrne.* London: William Pickering, [Printed by C. Whittingham at the Chiswick Press], 1847. RBK

Euler, Leonhard (1707–1783). *Methodus inveniendi lineas curvas . . . sive solutio problematis isoperimetrici.* Lausanne and Geneva: Marus-Michael Bousquet, 1744. SIBL; from the Astor Library

Fermat, Pierre de (1601–1665). *Varia opera mathematica.* Toulouse: Joannes Pech, 1679. SIBL; from the Astor Library

*LeClerc, Sébastien (1637–1714). *Practique de la géométrie sur le papier et sur le terrain.* Paris: Thomas Jolly, 1669. RBK [PAGE 31]

Lobachevski, Nikolai Ivanovich (1792–1856). *Géométrie imaginaire.* [Kasan], 1886. SIBL

Physics

Aldini, Giovanni (1762–1834). *An Account of the Late Improvements in Galvanism, with a series of curious and interesting experiments performed before the commissioner of the French National Institute, and lately repeated in anatomical theatres of London.* London: Cutrell and Gartin, 1803. PFORZHEIMER

Ampère, André Marie (1775–1836). "De l'action mutuelle de deux courans electriques" in: *Annales de chimie et de physique* (Paris), 2nd series, Vol. 15 (Sept.–Oct. 1820). WHEELER

Barlow, William (d. 1625). *Magneticall Advertisements: or, Divers Pertinent Observations.* London: Edward Griffin for Timothy Barlow, 1616. WHEELER

Borelli, Giovanni Alfonso (1608–1679). *De motu animalium.* Leyden: Petrus vander Aa, 1680–85. RBK

Chevreul, Michel Eugène (1786–1889). *Exposé d'un moyen de définir et de nommer les couleurs.* Paris: Firmin Didot Frères et Fils, 1861. GRD; from the Astor Library

Chladni, Ernst Florens Friedrich (1756–1827). *Entdeckungen über die Theorie des Klanges.* Leipzig: Weidmanns Erben und Reich, 1787. SIBL; from the Astor Library

*Descartes, René (1596–1650). *Discours de la méthode pour bien conduire sa raison, & chercher la verité dans les sciences. Plus la dioptrique. Les météores. Et la géométrie.* Leyden: Ian Maire, 1637. RBK [FRONTISPIECE]

*———. *Principia philosophiae.* Amsterdam: Louis Elzevir, 1644. RBK [PAGE 32]

*Fontana, Domenico (1543–1607). *Della trasportatione dell'Obelisco Vaticano.* Rome: Domenico Basa, 1590. PARSONS [PAGE 38]

Franklin, Benjamin (1706–1790). *Experiments and Observations on Electricity Made at Philadelphia in America.* London: E. Cave, 1751. BERG

Galvani, Luigi (1737–1798). *De viribus electricitatis in motu musculari.* Bologna: Typographia Instituti Scientiarum, 1791. WHEELER

Gernsback, Hugo (1884–1967). *The Wireless Telephone.* New York: Modern Electrics Publication, 1910. BURDEN

*Gilbert, William (1544–1603). *De magnete.* London: Peter Short, 1600. WHEELER [PAGE 35]

Goethe, Johann Wolfgang von (1749–1832). *Zur Farbenlehre.* Tübingen: J. G. Cotta, 1810. ROSIN

*Gorringe, Henry Honeychurch (1841–1885). *Egyptian Obelisks.* New York: for the author, 1882. STUART [PAGE 39]

Guericke, Otto von (1602–1686). *Experimenta nova (ut vocantur) Magdeburgica de vacuo spatio.* Amsterdam: Joannes Jansson à Waesberge, 1672. WHEELER

Hales, Stephen (1677–1761). *Vegetable Staticks, or, An Account of Some Statical Experiments on the Sap in Vegetables.* London: W. and J. Innys and T. Woodward, 1727. NYAM

Harris, Sir William Snow (1791–1867). *Observations on the Effects of Lightning on Floating Bodies*. London: G. and W. Nicol, 1823. WHEELER

Hauksbee, Francis (1666–1713). *Physico-Mechanical Experiments on Various Subjects*. London: R. Bruges, for the author, 1709. WHEELER

Hertz, Heinrich (1857–1894). *Untersuchungen über die Ausbreitung der elektrischen Kraft*. Leipzig: Johann Ambrosius Barth, 1892. SIBL

Huygens, Christiaan (1629–1695). *Horologium oscillatorium, sive, de motu pendulorum*. Paris: F. Muguet, 1673. SIBL; from the Astor Library

Kircher, Athanasius (1602–1680). *Magnes sive de arte magnetica . . . editio tertia*. Rome: Deversin & Masotta, 1654. WHEELER

The Landing of the French Atlantic Cable at Duxbury, Mass., July, 1869. Boston: Alfred Mudge & Son, 1869. PHOTOGRAPHY

Lana Terzi, Francesco (1631–1687). *Prodromo overo saggio di alcune inventioni nuove*. Brescia: Rizzardi, 1670. RBK

Latimer, Lewis Howard (1849–1928). *Incandescent Electric Lighting. A Practical Description of the Edison System*. New York: D. Van Nostrand Company, 1890. SC; signed by Arthur Schomburg, September 1914

Lilienthal, Otto (1848–1896). *Der Vogelflug als Grundlage der Fliegekunst*. Berlin: R. Gaertners, 1889. SIBL; from the Astor Library

Marat, Jean Paul (1743–1793). *Mémoires académiques, ou Nouvelles découvertes sur la lumière*. Paris: N. T. Méquignon, 1788. SIBL

Marum, Martin van (1750–1837). *Beschryving eener ongemeen groote electrizeer-machine, geplaatst in Teyvler's Museum te Haarlem . . . Description d'une très grande Machine Électrique*. Haarlem: Johann Enschedé & Son, 1785–87. WHEELER

Maxwell, James Clerk (1831–1879). *A Treatise on Electricity and Magnetism*. Oxford: Clarendon Press, 1873. WHEELER

Michelson, Albert Abraham (1852–1931). *Light Waves and Their Uses*. Chicago: The University of Chicago Press, 1903. SIBL

Musschenbroek, Peter van (1692–1761). *Physicae experimentales, et geometricae, de magnete*. Leyden: Samuel Luchtmans, 1729. WHEELER

Newton, Sir Isaac (1642–1727). *Opticks: or, A Treatise of the Reflexions, Refractions, Inflexions and Colours of Light*. London: Samuel Smith and Benjamin Walford, Printers to the Royal Society, 1704. SIBL

*———. *Philosophiae naturalis principia mathematica*. London: Joseph Streater, 1687. RBK [PAGE 36]

Nicholson, William (1735–1815). *An Introduction to Natural Philosophy*. London: J. Johnson, 1782. PFORZHEIMER

Norman, Robert (fl. 1590). *The Newe Attractive. Containing a Short Discourse of the Magnes or Loadstone*. London: E. Allde for Hew Astley, 1592. WHEELER

Ramelli, Agostino (ca. 1531–1608). *Le diverse et artificiose machine*. Paris: for the author, 1588. PARSONS; Bookplate of Nicolai Joseph Foucault

Ridley, Mark (1560–1624). *A Short Treatise of Magneticall Bodies and Motions*. London: Nicholas Okes, 1613. WHEELER

Switzer, Stephen (1682?–1745). *An Introduction to a General System of Hydrostaticks and Hydraulicks*. London: T. Astley, 1729. SIBL

Vail, Alfred (1807–1859). *Description of the American Electro Magnetic Telegraph: Now in operation between . . . Washington and Baltimore.* Washington, D.C.: J. & G. S. Gideon, 1845. WHEELER

Volta, Alessandro (1745–1827). *Collezione dell'opere del . . . Alessandro Volta.* Florence: Guglielmo Piatti, 1816. WHEELER

Winckler, Johann Heinrich (1703–1770). *Die Staerke der electrischen Kraft, des Wassers in glaesermen Gefaessen.* Leipzig: Bernhard Christoph Breitkopf, 1746. WHEELER

Young, Thomas (1773–1829). *The Bakerian Lecture. On the Theory of Light and Colours.* London, 1802. SIBL

[Playing Cards: Engineering]. London?, 18th century? PARSONS

[Playing Cards: Mechanics]. London: T. Tuttell, ca. 1700. PARSONS

Chemistry

Agricola, Georg (1494–1555). *De re metallica libri XII.* Basel: Hieronymus Froben, 1556. RBK; from the Astor Library

"The Alchemist," from the series *Lilliputian Humorists, Drawn as Big as Life.* London: Printed and Sold at the White Horse without Newgate, 1730. ARENTS:TOBACCO

Basilius Valentinus (fl. 15th century). *Revelation des mystères des teintures essencieles des sept metaux, et de leurs vertus medicinales . . . traduite par le Sieur Jean Israel.* Paris: Jacques de Senlecque & Jean Henault, 1646. SIBL

Becquerel, Henri (1852–1908). *Recherches sur une propriété nouvelle de la matière activité radiante spontanée ou radioactivité de la matière.* Paris: Firmin Didot et Cie, 1903. SIBL

Beguin, Jean (ca. 1550–ca. 1620). *Les élémens de chymie.* Lyon: Claude Chancey, 1665. SIBL

Boerhaave, Herman (1668–1738). *The Elements of Chemistry . . . Translated . . . by Timothy Dallowe.* London: J. and J. Pemberton, 1735. SIBL

*Boyle, Robert (1627–1691). *Nova experimenta physico-mechanica de vi aëris elastica.* Rotterdam: Arnold Leers Junior, 1669. WHEELER [PAGE 40]

Curie, Marie Sklodowska (1867–1934). *Traité de radioactivité.* Paris: Gauthier-Villars, 1910. SIBL

*———. *Untersuchungen über die radioaktiven Substanzen.* Brunschweig: Friedrich Viewig and Son, 1904. SIBL [PAGE 43]

Dalton, John (1766–1844). *A New System of Chemical Philosophy.* Manchester: S. Russell for R. Bickerstaff, London, 1808. SIBL

Faraday, Michael (1791–1867). *Experimental Researches in Chemistry and Physics.* London: Richard Taylor and William Francis, 1859. SIBL

Gillray, James (1757–1815). *"Scientific Researches! New Discoveries in Pneumatics! – or – an Experimental Lecture on the Power of Air."* London: Hannah Humphrey, May 23, 1802. PRINTS

Glauber, Johann Rudolf (1604–1668). *A Description of New Philosophical Furnaces, or, A New Art of Distilling.* London: Richard Coats for Thomas Williams, 1651. SIBL; Gift of Mrs. Henry Draper

Guyton de Morveau, Louis Bernard (1737–1816). *Méthode de nomenclature chimique.* Paris: Cuchet, 1787. SIBL

Haüy, René Just (1743–1822). *Traité de cristallographie . . . Atlas.* Paris: Bachelier & Huzard, 1822. SIBL

**Illustrierte Zeitung* (Leipzig), 107, no. 2773 (August 22, 1896), p. 231. GRD [PAGE 37]

Lavoisier, Antoine Laurent (1743–1794). *Elements of Chemistry, in a New Systematic Order . . . Translated . . . by Robert Kerr.* Edinburgh: William Creech, and sold in London by G. G. and J. J. Robinson, 1790. SIBL; from the Astor Library

———. *Nomenclature chimique . . . nouvelle édition*. Paris: Cuchet, 1789. SIBL

Lémery, Nicolas (1645–1715). *Cours de chymie . . . septième edition*. Paris: Estienne Michallet, 1690. RBK

Memorie della Società degli Spettroscopisti Italiani, Vol. II (1873). Palermo: Tipographico Lao. SIBL

Pantheo, Giovanni Agostino (fl. ca. 1517–35). *Ars et theoria transmutationis metallicae cum voarchadumia, proportionibus, numeris & iconibus rei accommodis illustrata*. Paris: V. Gaultherot, 1550. PARSONS

Parkinson, James (1755–1824). *The Chemical Pocket-Book*. Philadelphia: James Humphreys, 1802. SIBL

Pasteur, Louis (1822–1895). *Études sur le vin: ses maladies, causes qui les provoquent, procédés nouveaux*. Paris: Imprimerie Impériale, 1866. GRD

———. *Études sur le vinaigre, sa fabrication, ses maladies*. Paris: Gauthier-Villars and Victor Masson and Son, 1868. GRD; from the Astor Library

Priestley, Joseph (1733–1804). *Experiments and Observations on Different Kinds of Air . . . The second edition, corrected*. London: J. Johnson, 1774–77. SIBL

Scheele, Karl Wilhelm (1742–1786). *Chemical Observations and Experiments on Air and Fire . . . with a letter to him from Joseph Priestley. . . .* London: J. Johnson, 1780. SIBL

Watt, James (1736–1819). *Descriptions of a Pneumatic Apparatus . . . the second edition*. Birmingham: T. Pearson, 1795. SIBL

Medicine

Albinus, Bernhard Siegfried (1697–1770). *Tabulae sceleti et musculorum corporis humani*. London: H. Woodfall, 1749. NYAM

Anderson, Alexander (1775–1870). [Anatomical figure, male]. New York, 1799. PRINTS

Aselli, Gaspare (1581–1626). *De lactibus, sive lacteis venis*. Milan: Giovanni Battista Bidelli, 1627. NYAM

Bartisch, George (1535–1606). *[Ophthalmodouleia] das ist augendienst*. Dresden: Matthes Stöckel, 1583. NYAM

Beaumont, William (1785–1853). *Experiments and Observations on the Gastric Juice, and the Physiology of Digestion*. Plattsburgh, N.Y.: F. P. Allen, 1833. RBK

Bell, Charles (1774–1842). *The Anatomy of the Brain, explained in a series of engravings*. London: C. Whittingham for T. N. Longman and O. Rees, 1802. RBK; from the Astor Library

Bourgery, Jean Marc (1797–1849). *Traité complet de l'anatomie de l'homme*. Paris: C.-A. Delauney, 1831–54. GRD; from the Astor Library

*Bourneville, Desiré Magloire (1840–1909), and Paul Régnard (1850–1927). *Iconographie photographique de la Salpêtrière, Service de M. Charcot*. Paris: Bureau du Progrès Médical, V. Adrien Delahay et Cie, 1877–80. PHOTOGRAPHY [PAGE 48]

*Bravo, Francesco (fl. ca. 1553). *Opera medicinalia*. Mexico: Petrus Ocharte, 1570. RBK [PAGE 13]

*Cleyer, Andreas (1634–1697 or 1698). *Specimen medicinae sinicae, sive opuscula medica ad mentem sinensium*. Frankfurt: Johann Peter Zubrodt, 1682. GRD; from the Astor Library [PAGE 46]

Darwin, Charles (1809–1882). *The Expression of the Emotions in Man and Animals*. London: John Murray, 1872. PHOTOGRAPHY

Duchenne, Guillaume-Benjamin (1806–1875). *Physionomie humaine ou analyse électro-physiologique de l'expression des passions . . . Deuxième édition*. Paris: J.-B. Baillière et fils, 1876. PHOTOGRAPHY; from the Astor Library

Ehrlich, Paul (1854–1915). "On Immunity with Special Reference to Cell Life," pp. 424–48 in: *Proceedings of the Royal Society of London*, Vol. 66 (1900). SIBL

Eustachi, Bartolomeo (1520–1574). *Tabulae anatomicae.* Rome: Francesco Gonzaga, 1714. NYAM

Galton, Francis (1822–1911). *Finger Prints.* London and New York: Macmillan, 1892. GRD

*Gautier-d'Agoty, Arnaud Éloi (d. 1771). *Cours complet d'anatomie peint et gravé en couleurs naturelles.* Nancy: J. B. H. Leclerc, 1773. PRINTS [PLATE VI]

Geminus, Thomas (fl. 1540–60). *Compendiosa totius anatomie delineatio.* London: John Herford, 1545. RBK; John Shaw Billings Memorial Collection

———. *Compendiosa totius anatomie delineatio.* London: by the author, 1559. NYAM

*Genga, Bernardino (1620–1690). *Anatomia chirurgia, cioè istoria anatomica dell'ossa.* Rome: Nicolò Angelo Tinassi, 1672. NYAM [PAGE 47]

———. *Anatomia per uso et intelligenza del disegno ricercata non solo su gl'ossi e muscoli del corpo humano.* Rome: Domenico de Rossi, 1691. NYAM

Gillray, James (1757–1815). *"The Cow-Pock – or – The Wonderful Effects of the New Inoculation."* London: Hannah Humphrey, June 12, 1802. PRINTS

Gray, Henry (1825–1861). *Anatomy: Descriptive and Surgical . . . the drawings by H. V. Carter.* London: John W. Parker and Son, 1858. NYAM

*Harvey, William (1578–1657). *Exercitatio anatomica de motu cordis et sanguinis in animalibus.* Frankfurt: William Fitzer, 1628. RBK; from the Astor Library [PAGE 44]

Hunter, William (1718–1783). *Anatomia uteri humani gravidi tabulis illustrata . . . The Anatomy of the Human Gravid Uterus Exhibited in Figures.* Birmingham: John Baskerville, 1774. RBK; from the Astor Library

Jenner, Edward (1749–1823). *An Inquiry into the Causes and Effects of the Variolae Vaccinae.* London: Sampson Low for the author, 1798. NYAM

Kaempfer, Engelbert (1651–1716). *The History of Japan.* London: for the translator, 1727. ORT

Kaposi, Moriz (1837–1902). *Handatlas der Hautkrankheiten für studirende und Ärzte.* Vienna and Leipzig: Wilhelm Braumüller, 1898–1900. NYAM

Laënnec, René Théophile Hyacinthe (1781–1826). *De l'auscultation médiate, ou, Traité du diagnostic des maladies des poumons et du coeur.* Paris: J.-A. Brosson and J.-S. Chaudé, 1819. NYAM

Liebreich, Richard (1830–1917). *Atlas der Ophthalmoscopie. Darstellung des Augengrundes im gesunden und Krankhaften Zustande enthalten.* Berlin and Paris: Germer Baillière, 1863. NYAM

Mercuriale, Girolamo (1530–1606). *De arte gymnastica.* Venice: Guinta, 1573. NYAM

Smellie, William (1697–1763). *A Sett of Anatomical Tables, with explanations, and an abridgement of the practice of midwifery.* London: D. Wilson, 1754. NYAM

Snow, John (1813–1858). *On the Inhalation of the Vapour of Ether in Surgical Operations.* London: John Churchill, 1847. NYAM

Sömmerring, Samuel Thomas (1755–1830). *Icones embryonum humanorum.* Frankfurt am Main: Varrentrapp and Wenner, 1799. NYAM

Tagliacozzi, Gaspare (1545–1599). *De curtorum chirurgia per insitionem.* Venice: Gasparo Bindoni, the younger, 1597. RBK; from the Lenox Library

Virchow, Rudolf Ludwig Karl (1821–1902). *Die Cellularpathologie.* Berlin: August Hirschwald, 1858. NYAM

Willis, Thomas (1621–1675). *Cerebri anatome: cui accessit nervorum descriptio et usus.* London: James Flesher, for Joseph Martyn and John Allestry, 1664. NYAM

Natural History

*Agassiz, Louis (1807–1873). *Études sur les glaciers. . . . Dessinés d'après nature et lithographiés par Joseph Bettannier*. Neuchâtel: Jent & Gassman, à la Lithographie de H. Nicolet, 1840. SIBL [PAGE 50]

*Atkins, Anna (1799–1871). *Photographs of British Algae: Cyanotype Impressions*. Halstead Place, Sevenoaks, England: Anna Atkins, 1843–53. SPENCER; inscribed by the author to Sir John Herschel, Sevenoaks, October 1843 [PLATE VIII]

Audubon, John James (1785–1851). "Iceland or Jer Falcon." No. 13–2, Plate 19, from: *The Birds of America*. New York: Julius Bien, 1860–61. RBK; from the Lenox Library, Gift of Mrs. Henry Draper

Baer, Karl Ernst von (1792–1876). *De ovi mammalium et hominis genesi*. Leipzig: Leopold Voss, 1827. NYAM

Barton, Benjamin Smith (1766–1815). *Vegetable Materia Medica of the United States: or, Medical Botany*. Philadelphia: M. Carey & Son, 1817–19. ARENTS: BOOKS IN PARTS

Belon, Pierre (1517?-1564). *Portraits d'oyseaux, animaux, serpens, herbes, arbres, hommes et femmes, d'Arabie & Égypte*. Paris: Guillaume Cauellat, 1557. SPENCER

Bigelow, Jacob (1787–1879). *American Medical Botany, being a collection of the native medicinal plants of the United States*. Boston: Cummings and Hilliard, 1817–21. ARENTS: BOOKS IN PARTS

Bloch, Marcus Elieser (1723–1799). *Systema ichthyologiae iconibus*. Berlin: by the author, 1801. GRD

Darwin, Charles (1809–1882). *The Descent of Man, and Selection in Relation to Sex*. London: John Murray, 1871. BERG

———. *On the Origin of Species by Means of Natural Selection*. London: W. Clowes and Sons for John Murray, 1859. BERG

———. *The Zoology of the Voyage of H.M.S. Beagle*. London: Smith, Elder and Co., 1839–43. STUART

Deane, James (1801–1858). *Ichnographs from the Sandstone of Connecticut River*. Boston: Little, Brown & Company; London: S. Low, Son & Co., 1861. STUART

de Vries, Hugo (1848–1935). *Die Mutationstheorie*. Leipzig: Veit & Comp., 1901–1903. GRD

Durant, Charles Ferson (1805–1873). *Algae and Corallines of the Bay & Harbor of New York, illustrated with natural types*. New York: George P. Putnam, Printed by Narine & Co., 1850. RBK; Gift of Miss Emma Durant, 1904, daughter of the author

France. Commission des monuments d'Égypte. *Description de l'Égypte*. Paris: Imprimerie Impériale, 1808–28. ORT

Grew, Nehemiah (1641–1712). *The Anatomy of Plants*. London: W. Rawlins, for the author, 1682. SIBL; from the Astor Library

Halley, Edmund (1656–1742). *A New and Correct Chart of the Channel . . . 2d. ed.* London: Mount & Page, 1715. SLAUGHTER

Hamilton, Sir William (1730–1803). *Observation on Mount Vesuvius, Mount Etna, and Other Volcanos*. London: T. Cadell, 1772. PFORZHEIMER

*Hooke, Robert (1635–1703). *Micrographia, or, Some Physiological Descriptions of Minute Bodies Made by Magnifying Glasses*. London: John Martin and James Allestry for the Royal Society, 1665. RBK; from the Astor Library [PAGE 53]

Hough, Romeyn Beck (1857–1924). *The American Woods, exhibited by actual specimens and with copious explanatory text*. Lowville, N.Y.: by the author, 1888–1928. SCHLOSSER

*Imperato, Ferrante (1550–1625). *Dell'historia naturale*. Naples: Costantino Vitale, 1599. RBK [PAGE 52]

Leeuwenhoek, Antoni van (1632–1723). "A Specimen of Some Observations Made by a Microscope, contrived by M. Leewenhoeck in Holland, lately communicated by Dr. Regnerus De Graaf," in: *Transactions of the Royal Society* (London), Vol. VII (1672). SIBL; from the Astor Library

Leidy, Joseph (1823–1891). *Contributions to the Extinct Vertebrate Fauna of the Western Territories.* Washington, D.C.: Government Printing Office, 1873. STUART

Lonicer, Adam (1528–1586). *Naturalis historiae opus novum.* Frankfurt: Christian Egenolph, 1565. RBK

Ludwig, Christian Gottlieb (1709–1773). *Ectypa vegetabilium usibus medicis praecipue destinatorum et in pharmacopoliis.* Halle: J. G. Trampe, 1760–64. PRINTS

Lyell, Sir Charles (1797–1875). *Principles of Geology.* London: John Murray, 1830–33. SIBL

★ Merian, Maria Sibylla (1647–1717). *Metamorphosibus insectorum Surinamensium . . . Dissertation sur la generation et les transformations des insectes de Surinam.* The Hague: Pierre Gosse, 1726. RBK [PLATE VII]

Murchison, Sir Roderick Impey (1792–1871). *The Silurian System, Founded on Geological Researches.* London: John Murray, 1839. SIBL

Rondelet, Guillaume (1507–1566). *Libri de piscibus marinis.* Lyon: Matthew Bonhomme, 1554–55. RBK; from the Astor Library

Say, Thomas (1787–1834). *American Entomology, or Descriptions of the Insects of North America.* Philadelphia: Mitchell & Ames, William Brown, printer, 1817. RBK

———. *American Entomology, or Descriptions of the Insects of North America.* Philadelphia: Philadelphia Museum, published by Samuel Augustus Mitchell, William Brown, printer, 1824–28. RBK

Swammerdam, Jan (1637–1680). *Bybel der natuure . . . of historie der insecten.* Leyden: Isaak Severinus, 1737–38. WHEELER

Withering, William (1741–1799). *An Account of the Fox-glove, and Some of Its Medical Uses.* Birmingham: M. Swinney for G. G. J. and J. Robinson, London, 1785. NYAM

Worm, Ole (1588–1654). *Museum Wormianum.* Leyden: Elsevir, 1655. RBK; from the Astor Library

Illustration Techniques

Relief Printing

Woodcut

Liberale, Georgio, and Wolfgang Meyerpeck. Woodblock for *Bellis perennis.* Used in: Pietro Andrea Mattioli (1500–1577), *Commentarii in sex libros Pedacii Dioscoridis . . . de materia medica.* Prague: G. Melantrich, 1562, and later editions. MERTZ

———. *Bellis perennis.* London: Alecto Historical Editions, 1989. PRINTS; Gift of Virginia L. T. Gardner, V. L. T. Gardner Books, Santa Barbara, California

Mattioli, Pietro Andrea (1500–1577). *Commentarii in sex libros Pedacii Dioscoridis . . . de materia medica.* Venice: Valgrisiana, 1565. ARENTS: TOBACCO

Wood Engraving

Anderson, Alexander (1775–1870). *"Tees-Water Old or Unimproved Breed."* New York, ca. 1804. Wood-engraved block. PRINTS

———. *A General History of Quadrupeds. The figures engraved on wood, chiefly copied from the original of T. Bewick. First American edition, with an appendix containing some American animals not hitherto described.* New York: G. & R. Waite, 1804. DUYCKINCK

Anthony, A. V. S. (1835–1906). Woodblock with drawing of landscape, prepared for engraving but uncut. PRINTS; Gift of his daughter, Mrs. Henry Perkins, 1924

Wooden frame to hold block for engraving used by Anthony. PRINTS; Gift of his daughter, Mrs. Henry Perkins, 1924

Gravers used by Anthony. PRINTS; Gift of his daughter, Mrs. Henry Perkins, 1924

Intaglio Printing

Line Engraving

Cheselden, William (1688–1752). *Osteographia, or The Anatomy of the Bones*. London: W. Bowyer for the Author, 1733. NYAM

Cruikshank, George (1792–1878). [*Marcel's Last Minuet*]. [London, 1838]. Engraved and etched steel plate. PRINTS; John Shaw Billings Memorial Collection.

Durand, Asher B. (1796–1886). *"Mr. Cowell as Crack."* [Philadelphia, 1826]. Engraved copper plate. PRINTS; Gift of his daughter, Alice Burt

———. *"Mr. Cowell as Crack."* Philadelphia: A. R. Poole, 1826. PRINTS; Gift of his daughter, Alice Burt

Engraving pillow, loop, and dauber used by Edwin Davis French (1851–1906). PRINTS; Gift of Mrs. E. D. French

Gravers, scraper, needles, and ivory caliper. PRINTS

Mezzotint

Gole, Jacob (1660–1737). *"America,"* from the series *The Continents, Represented by Ladies in Costumes*. Amsterdam: J. Gole, n.d. PRINTS; Gift of Edward G. Kennedy

Roulette tools used by A. V. S. Anthony (1835–1906) and Edwin Davis French (1851–1906), and ground tool used by French. PRINTS; Gift of Mrs. Edwin Davis French

Planographic Printing

Lithography

Hullmandel, Charles Joseph (1789–1850). *The Art of Drawing on Stone*. London: C. Hullmandel & R. Ackermann, [1824]. PRINTS

*Lear, Edward (1812–1888). *Illustrations of the Family of Psittacidae, or Parrots*. London: by the author, 1832. STUART [PLATE IX]

Poulbot, Francisque (1879–1946). [Holiday card for Bella C. Landauer, 1928]. N.p., [1928]. Lithography stone prepared with drawing. PRINTS; Gift of Bella C. Landauer

———. "Joyeux Noël, Bella C. Landauer." [Paris?, 1928]. PRINTS; Gift of Bella C. Landauer

Chromolithography

Prang, Louis (1824–1909). *Prang's Prize Babies: How This Picture Is Made*. Boston: L. Prang & Co., [1888]. PRINTS

Photography

Daguerre, Louis Jacques Mandé (1789–1851). *Historique et description des procédés du daguerréotype et du diorama*. Paris: Alphonse Giroux et Cie, 1839. RBK; John Shaw Billings Memorial Collection

Lick Observatory, University of California. *Transparencies of the Moon*. [Mount Hamilton, Calif., ca. 1896]. PHOTOGRAPHY; Gift of Prof. Edward S. Holden

Nasmyth, James Hall (1808–1890), and James Carpenter (1840–1899). *The Moon: considered as A Planet, A World, and A Satellite . . . second edition*. London: J. Murray, 1874. SIBL; Gift of Mrs. Henry Draper.

United States. War Department. Surgeon General's Office. Army Medical Museum. "Seven Adult Male Sandwich Islanders" and "Taking Composite Negative of Skulls" from: *Composite Photographs of Crania*. Washington, D.C.: Army Medical Museum, [1884]. PHOTOGRAPHY; Gift of John Shaw Billings

Vogel, Hermann Wilhelm (1834–1898). *La photographie et la chemie de la lumière*. Paris: Ballière, 1876. PHOTOGRAPHY

Selected Bibliography

American Institute of Electrical Engineers. Library. *Catalogue of the Wheeler Gift of Books, Pamphlets and Periodicals in the Library of the American Institute of Electrical Engineers*. New York: The Institute, 1909.

Arber, Agnes. *Herbals, Their Origin and Evolution: A Chapter in the History of Botany, 1470–1670*. Cambridge, England: Cambridge University Press, 1938.

Atkins, Anna. *Sun Gardens. Victorian Photograms.* Text by Larry J. Schaaf. New York: Hans P. Kraus, Jr., An Aperture Book, 1985.

Blunt, Wilfrid, and William T. Stearn. *The Art of Botanical Illustration*. Woodbridge, Suffolk: Antique Collectors' Club in association with The Royal Botanic Gardens, Kew, 1994.

Bruno, Leonard C. *The Tradition of Science: Landmarks of Western Science in the Collections of the Library of Congress*. Washington, D.C.: Library of Congress, 1987.

Butterfield, Herbert. *The Origins of Modern Science, 1300–1800*. Revised edition. New York: The Free Press, Macmillan, 1965.

Bynum, W. F.; E. J. Browne; and Roy Porter. *Dictionary of the History of Science*. Princeton, N.J.: Princeton University Press, 1984.

Carter, John; Percy H. Muir; and others. *Printing and the Mind of Man: A Descriptive Catalogue Illustrating the Impact of Print on the Evolution of Western Civilization*. Second edition, revised and enlarged. Munich: Karl Pressler, 1983.

Cave, Roderick, and Geoffrey Wakeman. *Typographia naturalis [A History of Nature Printing]*. Wymondham, Leicestershire: Brewhouse Press, 1967.

Choulant, Ludwig. *History and Bibliography of Anatomic Illustration*. Translated and edited by Mortimer Frank. Chicago: The University of Chicago Press, 1920.

Christie's, New York. *The Haskell F. Norman Library of Science and Medicine*. 3 vols. New York: Christie's Inc., 1998.

Cohen, I. Bernard. *Revolution in Science*. Cambridge, Mass., and London: The Belknap Press of Harvard University Press, 1985.

Dibner, Bern. *Heralds of Science as Represented by Two Hundred Epochal Books and Pamphlets in the Dibner Library*. Norwalk, Conn.: Burndy Library, 1955.

Ford, Brian J. *Images of Science: A History of Scientific Illustration*. London: The British Library, 1992.

Garrison, Fielding H. *An Introduction to the History of Medicine*. Fourth Edition. Philadelphia and London: W. B. Saunders Company, 1929; reprinted 1960.

Gascoigne, Bamber. *How to Identify Prints: A Complete Guide to Manual and Mechanical Processes from Woodcut to Ink-jet*. London: Thames and Hudson, 1986.

Grolier Club. *One Hundred Books Famous in Medicine*. New York: The Grolier Club, 1995.

Grolier Club. *One Hundred Books Famous in Science . . . by Harrison D. Horblit*. New York: The Grolier Club, 1964.

Gross, Miriam T. "Classic Illustrated Zoologies (1550–1900) in the Research Collections of The New York Public Library: A Select Bibliography with Commentaries," *Biblion: The Bulletin of The New York Public Library*, 2, no. 2 (Spring 1994): 19–123.

Johns, Adrian. *The Nature of the Book: Print and Knowledge in the Making*. Chicago and London: University of Chicago Press, 1998.

Linda Hall Library. University of Missouri. *Out of This World: The Golden Age of the Celestial Atlas*. The online catalogue (www.lhl.lib.mo.us), written by William B. Ashworth, Jr., for an exhibition presented at the Linda Hall Library in November 1995–February 1996. Information on ordering a printed version of the catalogue is available on the website.

The New York Academy of Medicine. *Author Catalog of the Library*. 43 vols. Boston: G. K. Hall, 1969.

The New York Public Library. *The Animal Illustrated, 1550–1900*. Text by Joseph Kastner, with commentaries by Miriam T. Gross. New York: Harry N. Abrams, 1991.

The New York Public Library. *The Bird Illustrated, 1550–1900*. Text by Joseph Kastner, with commentaries by Miriam T. Gross. New York: Harry N. Abrams, 1988.

The New York Public Library. *Nature Illustrated, 1550–1900*. Text by Bernard McTigue. New York: Harry N. Abrams, 1989.

Partington, James R. *A History of Chemistry*. 4 vols. London: Macmillan; New York: St. Martin's Press, 1961–70.

Picturing Knowledge: Historical and Philosophical Problems Concerning the Use of Art in Science. Edited by Brian S. Baigrie. Toronto and Buffalo: University of Toronto Press, 1996.

Reese, William S. *Stamped with a National Character: Nineteenth Century American Color Plate Books*. New York: The Grolier Club, 1999.

Robin, Harry. *The Scientific Image from Cave to Computer*. New York: Harry N. Abrams, Inc., 1992.

Royal Academy of Arts and The Pierpont Morgan Library. *The Painted Page: Italian Renaissance Book Illumination, 1450–1550*. Munich and New York: Prestel, 1994.

Schmitz, Barbara. *Islamic Manuscripts in The New York Public Library*. New York and Oxford: Oxford University Press and The New York Public Library, 1992.

Sudhoff, Karl. *The Earliest Printed Literature on Syphilis*. Florence: R. Lier & Co., 1925.

Tufte, Edward. *Envisioning Information*. Cheshire, Conn.: Graphics Press, 1990.

Twyman, Michael. *Lithography, 1800–1850*. London, New York, and Toronto: Oxford University Press, 1970.

Vesalius, Andreas. *The Illustrations from the Works of Andreas Vesalius of Brussels*. New York: Dover Publications, 1973.

Acknowledgments

A PROJECT OF the size and complexity of *Seeing Is Believing* – both this book and the exhibition it accompanies – could not have succeeded without encouragement, support, and input from many others, particularly our colleagues at the Library. Special thanks to all those who worked with us to ensure the success of the exhibition and this book.

For the use in the exhibition of important materials from the collections of The New York Academy of Medicine, we extend our thanks to Edward T. Morman, Associate Academy Librarian of Historical Collections and Programs; Lois Fischer Black, Curator of Rare Books and Manuscripts; Lilli Sentz; and Caroline Duroselle-Melish. Thanks also to the Academy's Conservation Department, headed by Elaine Schlefer.

For the loan of the presentation copy of Vesalius's *De humani corporis fabrica* (1543), sold at Christie's on March 18, 1998, we are indebted to Francis Wahlgren, who first suggested The New York Public Library as a place where the anonymous purchaser might display the book. Patrick Cooney, representing the lender, was ever helpful in seeing to the loan. Thanks also to John F. Reed, Director of The LuEsther T. Mertz Library, The New York Botanical Garden, for the loan of an early herbal woodblock. Lynn Gamwell, Director of the Binghamton University Art Museum and curator of the exhibition *Beyond Appearances: Imagery in Science at the Millennium*, provided information and inspiration.

And, perhaps most importantly, for support on the home front we thank Michael and Nicolas Mirabile, and Paul Helfer.

Captions for Color Plates

PLATE I: Sacro Bosco's astronomy textbook *De sphaera*, based on classical astronomy, had established by 1220 that the Earth was a sphere. This illustration of the medieval cosmos, from a manuscript copy of Sacro Bosco's work (ca. 1275), is characteristic of the illustrations that persisted until well into the seventeenth century.

PLATES II–III: A professor of mathematics at the University of Ingolstadt, Apianus was also one of the most successful popularizers of astronomy and geography in the sixteenth century. For his *Astronomicum caesareum* (1540), he created thirty-five elaborate mechanical devices, called volvelles, featuring multiple disks, pointers, and string, that could be used to easily determine the position and movement of the planets, times of ellipses, and other astronomical phenomena.

PLATE IV: In the dedication copy of Vesalius's *De humani corporis fabrica*, presented to the Holy Roman Emperor, Charles V, in the autumn of 1543, all the illustrations were hand-colored, most likely under Vesalius's supervision, with highlights in liquid gold and silver. As the title-page illustration, showing the author pointing to a part of the female cadaver on the table in front of him, makes clear, Vesalius insisted that physicians do their own dissecting and not leave the work to assistants.

PLATE V: E. L. Trouvelot, a French-born artist and amateur astronomer, spent 1872–74 working with the 15-inch refractor at the Harvard Observatory. During this time he prepared a number of large pastel drawings representing, as he wrote in his *Astronomical Drawings* (1882), "the celestial phenomena as they appear to the trained eye and to an experienced draughtsman through the great modern telescopes." Fifteen of these drawings were reproduced using chromolithography, an illustration process that was at the zenith of its development in the 1880s. This one shows the Aurora Borealis, as observed on March 1, 1872.

PLATE VI: Arnaud Éloi Gautier-d'Agoty's *Cours complet d'anatomie* (1773) was one of the few anatomical books illustrated using the mezzotint process; medical texts generally required greater fidelity and precision of detail than could be achieved with mezzotint. The last survivors of the medieval manuscript tradition of illustrations depicting skeletons, musclemen, as well as the muscles of the face and eye (shown here), Gautier-d'Agoty's life-size paintings are striking for their artistic power, rather than their anatomical usefulness.

PLATE VII: Both an author and an illustrator, Maria Sibylla Merian was one of the world's great entomologists. Her first important work, *Wondrous Transformation of Caterpillars* (1679), included 186 European moths, butterflies, and other insects, all drawn from life rather than from preserved specimens. Her greatest work, *On the Metamorphosis of the Insects of Surinam*, followed a similar format. The first edition appeared in Latin and Dutch in 1705; subsequent editions included those in Latin alone, in Dutch alone, and in Latin and French; this illustration is reproduced from the latter edition, published in 1726.

PLATE VIII: Anna Atkins was the first woman photographer, and the first person to print and publish her own book illustrated entirely by photography. Published in parts on a regular schedule over ten years (1843–53), in an edition of probably not many more than the dozen copies known today, her *British Algae* stands as an important and generally overlooked milestone in the history of scientific illustration.

PLATE IX: Best known today for his "nonsense books," created after he began to lose his eyesight, Edward Lear was one of the finest bird painters. His first project, *Illustrations of the Family of Psittacidae, or Parrots* (1832) was also the first illustrated work of ornithology devoted to a single family of birds, and the first English bird folio book to be produced using lithographed plates that were then colored by hand. His *Parrots* established a format and style that was made famous by John Gould, for whom Lear worked following its publication.

SUPPORT FOR this exhibition and The New York Public Library's Exhibitions Program has been provided by Pinewood Foundation. This exhibition has also been made possible by funding from The Pfizer Foundation, Inc. Additional support for this exhibition has been provided by a grant from the New York Council for the Humanities, a state program of the National Endowment for the Humanities, as part of State Humanities Month.

THIS PUBLICATION is made possible by The Bertha and Isaac Liberman Foundation, Inc. in memory of Ruth and Seymour Klein.